SpringerBriefs in Petroleum Geoscience & Engineering

The SpringerBriefs series in Petroleum Geoscience & Engineering promotes and expedites the dissemination of substantive new research results, state-of-the-art subject reviews and tutorial overviews in the field of petroleum· exploration, petroleum engineering and production technology. The subject focus is on upstream exploration and production, subsurface geoscience and engineering. These concise summaries (50–125 pages) will include cutting-edge research, analytical methods, advanced modelling techniques and practical applications. Coverage will extend to all theoretical and applied aspects of the field, including traditional drilling, shale-gas fracking, deepwater sedimentology, seismic exploration, pore-flow modelling and petroleum economics. Topics include but are not limited to:

- Petroleum Geology & Geophysics
- Exploration: Conventional and Unconventional
- Seismic Interpretation
- Formation Evaluation (well logging)
- Drilling and Completion
- Hydraulic Fracturing
- Geomechanics
- Reservoir Simulation and Modelling
- Flow in Porous Media: from nano- to field-scale
- Reservoir Engineering
- Production Engineering
- Well Engineering; Design, Decommissioning and Abandonment
- Petroleum Systems; Instrumentation and Control
- Flow Assurance, Mineral Scale & Hydrates
- Reservoir and Well Intervention
- Reservoir Stimulation
- Oilfield Chemistry
- Risk and Uncertainty
- Petroleum Economics and Energy Policy

Contributions to the series can be made by submitting a proposal to the responsible Springer contact, Charlotte Cross at charlotte.cross@springer.com or the Academic Series Editor, Prof Dorrik Stow at dorrik.stow@pet.hw.ac.uk.

More information about this series at http://www.springer.com/series/15391

Ehsan Khamehchi · Mohammad Reza Mahdiani

Gas Allocation Optimization Methods in Artificial Gas Lift

 Springer

Ehsan Khamehchi
Faculty of Petroleum Engineering
Amirkabir University of Technology
 (Tehran Polytechnic)
Tehran
Iran

Mohammad Reza Mahdiani
Faculty of Petroleum Engineering
Amirkabir University of Technology
 (Tehran Polytechnic)
Tehran
Iran

ISSN 2509-3126   ISSN 2509-3134 (electronic)
SpringerBriefs in Petroleum Geoscience & Engineering
ISBN 978-3-319-51450-5   ISBN 978-3-319-51451-2   (eBook)
DOI 10.1007/978-3-319-51451-2

Library of Congress Control Number: 2016961272

Printed on acid-free paper

This Springer imprint is published by Springer Nature
The registered company is Springer International Publishing AG
The registered company address is: Gewerbestrasse 11, 6330 Cham, Switzerland

This book is dedicated to everyone who is interested in petroleum engineering

Preface

Using optimization methods is of great importance in different aspects of petroleum engineering, including production engineering and gas lift. Gas allocation optimization is crucial in maximizing the gas lift performance. There are different topics on gas allocation optimization, which are necessary for production engineers to know, such as the fitness function, the constraints, etc. Thus, we wrote this book and included different points on gas allocation optimization. Here, different methods for modeling the problem, distinct optimization constraints, and various optimizers have been discussed.

Studying this book is recommended to engineers and students who are interested in gas lift optimization.

Tehran, Iran

Ehsan Khamehchi
Mohammad Reza Mahdiani

Contents

Abbreviations

Ai	Injection port size, ft^2
B_g	FVF of gas at injection point
C_g	Cost of gas lift $/Mscf
D_i	Injection depth, ft
D_t	Tubing depth, ft
D_{well}	Well depth, ft
E	Orifice efficiency factor, 0.9
F_1, F_2	Asheim stability factors
f_o	Oil fraction
g	Acceleration of gravity, ft/s^2
g^k	The gradient of "f" at Q_g^k
GLR	Gas liquid ratio, SCF/STB
ID_c	Casing inner diameter, in
ID_t	Tubing inner diameter, in
IFT	Surface tension, dyne/cm
J	Productivity index, scf/s.psi
k	Counter of iterations
M_a	Apparent molecular weight
m	Slope in equal slope method
OD_t	Tubing outer diameter, in
Orifice size	Orifice size, 1/64 in
P	Net profit of oil $/bbl
P*	Pressure reservoir from well test
P_b	Bubble point pressure, psi
PI	Productivity index, STB/day/psi
P_{pr}	Pseudo reduced pressure
P_R	Reservoir pressure, psi
p_{ti}	Tubing flow pressure at gas injection point, psi
P_{wh}	Well head pressure, psi
q_{fi}	Flow rate of reservoir fluids at injection point, ft^3/s

Q_g	Injected gas, MMSCF/day
q_{gi}	Injection rate of each well
q_{gi}	Flow rate of lift gas at injection point, ft^3/s
q_{gtotal}	Total available lift gas
q_{lsc}	Flow rate of liquids at standard conditions, scf/s
Q_O	Total amount of oil production
Q_o	Produced oil of each well, STB/day
Q_t	Total produced oil, STB/day
T_{pr}	Pseudo reduced temperature
T_R	Reservoir temperature, F
T_{wh}	Well head temperature
V_C	Gas conduit volume
V_t	Tubing volume downstream of gas injection point, ft^3
WC	Water cut, %
α	Step length in numerical optimization
γ_g	Gas gravity
γ_{ginj}	Injection gas gravity
γ_w	Water gravity
μ_o	Oil viscosity, cp
ρ	Oil gravity, API
ρ_{fi}	Reservoir fluid density at injection point, lbm/ft^3
ρ_g	Gas density, lbm/ft^3
ρ_{gi}	Lift-gas density at the injection point, lbm/ft^3
ρ_{gsc}	Lift-gas density at standard surface conditions, lbm/scf

Chapter 1
An Introduction to Gas Lift

Abstract When the reservoir pressure declines, the production oil rate decreases and falls below the economic limit. One of the methods to increase the production rate is the gas lift. In this process, gas is injected to the well from the annulus. Then at the injection point it enters the tubing and is dissolved in the tubing's fluid which causes the reduction of the fluid density. Thus, the head pressure of the fluid column decreases, and then the production rate increases. Usually in gas lift projects there is a limited amount of gas that should be allocated between some wells in a way that some limitations such as the amount of fluid production, injection rate and facilities constraints are satisfied. Different wells have their specific properties and thus their respond to the injected gas is different. Finding a method regarding these constraints that maximizes the production is the subject of gas allocation optimization. This method should consider the properties of the wells, reservoir and facilities, as well as its proposed optimum point and should satisfy all constraints. In this chapter the physics of the gas lift and a briefing on how to formulate the gas allocation optimization problems will be given.

Keywords Constraint optimization · Gas injection · Gas lift · Gas allocation · Optimization methods

1.1 Introduction

As the production continues, reservoir pressure declines and causes a reduction in the petroleum production rate. In these cases, using artificial lift methods such as gas lift is inevitable (Shen et al. 2013). Gas lift helps the reservoir which is able to drive oil to the well bottom but not to the surface to produce in an economic rate (Mahdiani and Khamehchi 2015b). In this method, gas is injected to the well, dissolves in oil and decreases the head pressure in the well. Thus, the pressure difference between the bottom hole and the well head increases and causes an increment in oil production (Khamehchi et al. 2009). Figure 1.1 shows a schematic of this operation.

© The Author(s) 2017
E. Khamehchi and M.R. Mahdiani, *Gas Allocation Optimization Methods in Artificial Gas Lift*, SpringerBriefs in Petroleum Geoscience & Engineering, DOI 10.1007/978-3-319-51451-2_1

Fig. 1.1 A schematic of gas lift operation (Ghassemzadeh et al. 2015)

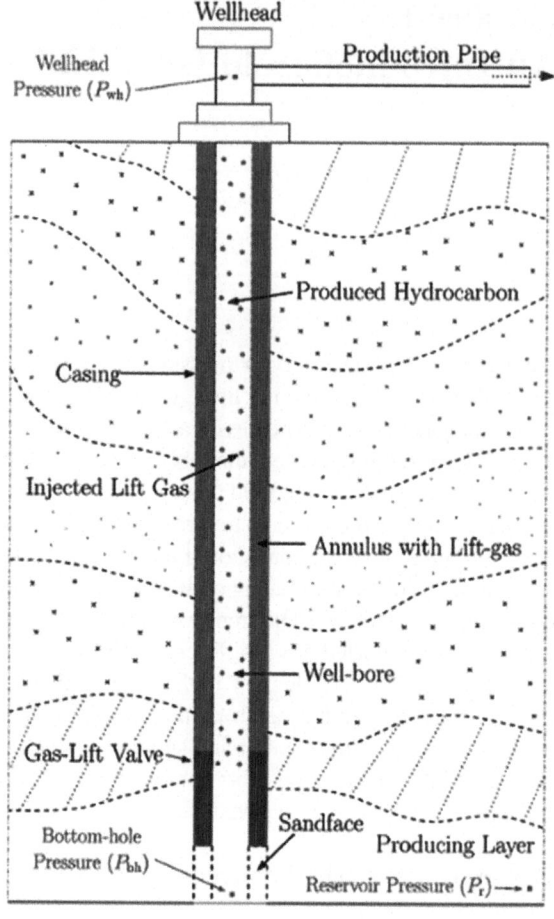

In the formulation of every optimization problem, first a fitness function should be defined. The fitness function catches some inputs (here gas injection rate (q_{gi}) of different wells) and returns the output. Output can be a single value or an array of values (multi objective) (Deb 2001; Zhu 2015; Neustadt 2015). The purpose of the optimizer is to maximize or minimize the output. The relation between the outputs and inputs is called the fitness or objective function (Rao and Rao 2009; Bejan 1995; Newman 2008). In addition, it is very common for every optimization problem to have some constraints (Woldesenbet 2007; Takahama and Sakai 2009; Bhatti 2000). The optimizer should find a point which minimizes or maximizes the output while satisfying all constraints.

The increment of the oil depends on many factors such as gas injection rate and reservoir and well properties (Chithra Chakra et al. 2015; Jacoud et al. 2015). In addition, greater oil production does not necessarily mean more profit, and the cost of production, such as the cost of the compressors, should be considered. Usually, in gas allocation most of the involved parameters cannot be changed due to previous design and installation and only the gas injection rates remain as changeable, and thus the maximum profit corresponds to the maximum oil production (Takács 2005). Gas allocation optimization is a process of finding the best allocation which causes the maximum profit (Sifuentes et al. 1996). Figure 1.2 shows a schematic of the gas allocation process.

This optimization based on the performance of each well, assigns some gas to each well in which the total profit of all wells is maximized and also some constraints are satisfied (Mahdiani and Khamehchi 2015a).

In this book, different aspects of gas allocation optimization will be discussed. First, the formulation of the problem and different methods (such as the building the proxy models) will be reviewed and the advantages and disadvantages of each method will be surveyed. Afterwards, different constraints that can be considered in this problem will be discussed and finally the optimization algorithms and their efficiency will be reviewed.

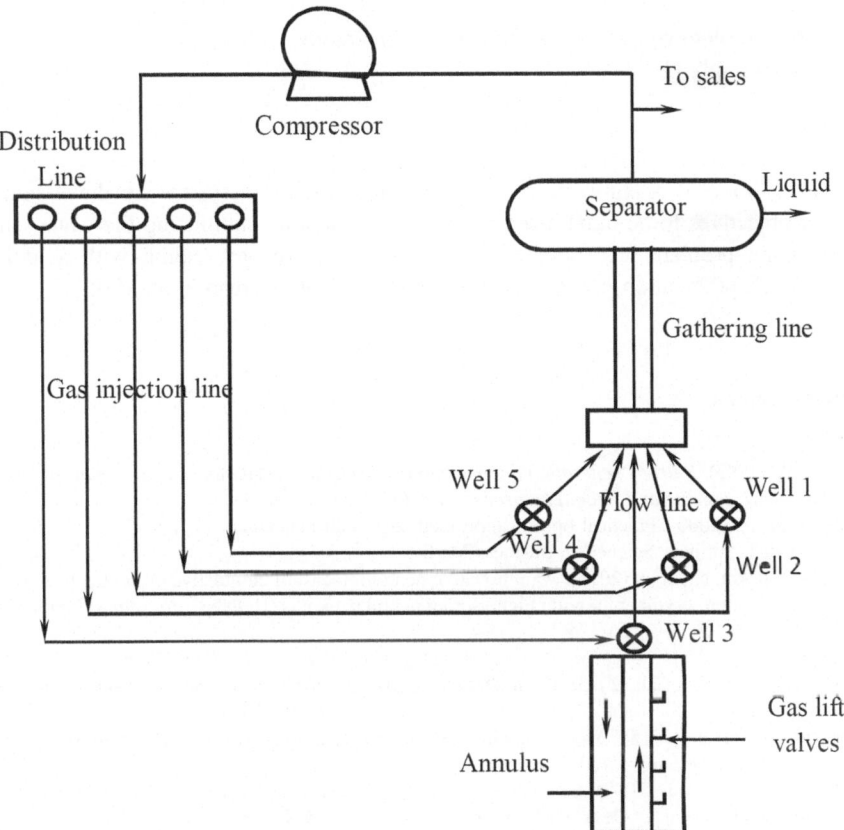

Fig. 1.2 A schematic of gas allocation (Monfared and Helalizadeh 2013)

1.2   Formulation of the Problem

The problem is that the total production oil rate (Q_O) which is the sum of individual production of each well (q_{Oi}) should be maximized by allocating an optimum arrangement of gas injection rate (q_{gi}) to the wells. In mathematical form it can be expressed as (1.1) and (1.2) (Nishikiori et al. 1989):

$$Q_O = \sum_{i=1}^{n} q_{oi} = f(q_{g1} \cdot q_{g2} \ldots \ldots q_{gn}) = f(Q_g) \tag{1.1}$$

$$Q_g = \left(q_{g1} \cdot q_{g2} \ldots \ldots q_{gn}\right)^{T} \tag{1.2}$$

The superscript T denotes the transposition of the matrix. Based on the above equation, the gas allocation optimization problem can be expressed as (1.3):

$$MaxQ_o = Maxf(Q_g) \tag{1.3}$$

This problem can be exposed to some constraints.
For example (1.4):

$$q_{gi} \geq 0 \quad for \, i = 1.2 \ldots . . n \tag{1.4}$$

In addition to maximizing the total oil production, the system can have some other objectives to be maximized or minimized such as minimizing the water cut. When the problem has more than one objective, we are dealing with a multi objective optimization problem (Sifuentes et al. 1996; Osman et al. 2005).

References

Bejan, A. (1995). *Entropy generation minimization: The method of thermodynamic optimization of finite-size systems and finite-time processes.* CRC Press.

Bhatti, M. A. (2000). Practical optimization methods: With mathematica applications: CD-ROM included. Springer Science & Business Media.

Chithra Chakra, N., et al. (2015). An innovative neural forecast of cumulative oil production from a petroleum reservoir employing higher-order neural networks (HONNs). *Journal of Petroleum Science and Engineering, 26*(3–4), 18–33.

Deb, K. (2001). *Multi-objective optimization using evolutionary algorithms.* Wiley.

Ghassemzadeh, S., et al. (2015). Optimization of gas lift allocation for improved oil production under facilities constraints. *Geosystem Engineering, 5*(3), 39–47.

Jacoud, A., et al. (2015). Modelling and extremum seeking control of gas lifted oil wells. *IFAC-PapersOnLine, 48*(2012), 21–26.

Khamehchi, E., et al. (2009). Continuous gas lift optimization with a novel genetic algorithm. *Australian Journal of Basic and Applied Sciences, 1*(4), 587–594.

Mahdiani, M. R., & Khamehchi, E. (2015a). Preventing instability phenomenon in gas-lift optimization. *Iranian Journal of Oil & Gas Science and Technology, 4*(1), 49–65.

Mahdiani, M. R., & Khamehchi, E. (2015b). Stabilizing gas lift optimization with different amounts of available lift gas. *Journal of Natural Gas Science and Engineering, 26,* 18–27.

Monfared, M., & Helalizadeh, A. (2013). Simulation and gas allocation optimization of gas lift system using genetic algorithm method in one of Iranian oil field to sales distribution liquid gathering line gas injection line valves Flow line well 2. *Journal of Basic and Applied Scientific Research, 3*(3), 732–738.

Neustadt, L. W. (2015). *Optimization: a theory of necessary conditions.* Princeton University Press.

Newman, T. R. (2008). *Multiple objective fitness functions for cognitive radio adaptation.* ProQuest.

Nishikiori, N., et al. (1989). An improved method for gas lift allocation optimization. In *SPE Annual Technical Conference and Exhibition.* Society of Petroleum Engineers.

Osman, M. S., Abo-Sinna, M. A., & Mousa, A. A. (2005). An effective genetic algorithm approach to multiobjective resource allocation problems (MORAPs). *Applied Mathematics and Computation, 163*(2), 755–768.

Rao, S. S., & Rao, S. S. (2009). *Engineering optimization: Theory and practice.* Wiley.

Shen, Z., et al. (2013). Artificial lift technique of multistage sliding vane pump used in thermal production well. *Petroleum Exploration and Development, 40*(5), 651–655.

Sifuentes, W., et al. (1996). An improved method for gas lift allocation optimization. *Journal of Natural Gas Science and Engineering, 48*(3), 1–5.

Takács, G. (2005). *Gas lift manual.* PennWell.

Takahama, T., & Sakai, S. (2009). *Constraint-handling in evolutionary optimization.* Springer.

Woldesenbet, Y. (2007). *Uncertainty and constraint handling in evolutionary algorithms.* ProQuest.

Zhu, J. (2015). *Optimization of power system operation.* Wiley.

Chapter 2
The Fitness Function of Gas Allocation Optimization

Abstract Fitness function is the heart of optimization problems. In gas allocation optimization, the fitness function takes the injection rates of different wells and returns the total revenue, or in some cases the total production oil rate. If the production rate of a well for different amount of injection rate could be calculated, then the fitness function has been found. There are different methods to gain the fitness function, one is the use of nodal analysis in which the well length is divided to some sections in order to ensure the small change of pressure and temperature and thus almost constant pvt properties of the fluid in the section length. Afterwards, using the empirical correlations, the production oil rate for a specific injected gas rate is calculated. This method can be done by the analytical approach, using equations such as mass balance, momentum balance, etc. Another method for creating a fitness function is using proxy models there are different methods to create the proxy models and they are relatively fast but their problem is their low accuracy. The mentioned methods can calculate the oil rate, but if the net profit is required, it can be gained using the economic methods in addition to the calculated production rates. The final point is that during the production life some economic and technical parameters change. As an example of a technical one, the reservoir condition is time dependent and thus there is a need to involve that in long term problems and here, the need for integrated modeling discloses. In this chapter, all the mentioned topics will be discussed in more details.

Keywords Nodal analysis · Proxy models · Analytical modeling · Integrated modeling

2.1 Introduction

Similar to other optimization problems, gas allocation optimization also needs to have a fitness function. This function takes the input parameters and returns the desired value (profit or rate of production) (Liu et al. 2015; Yang et al. 2016).

© The Author(s) 2017   7
E. Khamehchi and M.R. Mahdiani, *Gas Allocation Optimization Methods in Artificial Gas Lift*, SpringerBriefs in Petroleum Geoscience & Engineering, DOI 10.1007/978-3-319-51451-2_2

The purpose of the optimization is to maximize this function. In previous works, there were some methods to calculate this.

2.2   Technical Factors

If optimization methods are used in the design stage, different parameters such as tubing diameter, injection depth and compressor of other facilities can be optimized (de Souza et al. 2010). The gas allocation optimization is usually used when the design stage is completed and there is no possibility to change the well parameters. Thus, the only parameter that can be changed is the injection rate.

2.2.1   Nodal Analysis

In this method, for a supposed production rate, the well is divided to some sections in which the change of pressure and temperate is so small that its effects on the pvt properties of the fluid will be insignificant. Then starting from the uppermost section and using multiphase flow models, and an estimation of the pvt properties with different correlations (Asoodeh and Bagheripour 2012; Mohsen-Nia 2014; Asadisaghandi and Tahmasebi 2011) the pressure and temperature at the end of the section are estimated. Again using the corrected pvt properties and repeating the above procedure, the pressure and temperature at the end of the section is calculated. The pvt properties that are needed to be estimated at different pressure and temperature in nodal analysis studies are critical temperature and pressure, viscosity, gas compressibility factor and solution gas oil ratio. There are different correlations in the literature for estimating these parameters, for example for critical properties there are Klincewicz and Reid (1984), Whitson (1984), Sutton (1985), Joback and Reid (1987), Avaullee et al. (1997), and Li et al. (2016). One of the most common is the Sutton method in which its equation is as follows:

$$p_{pc}(psi) = 756.8 - 131\gamma_g - 3.6\gamma_g^2 \tag{2.1}$$

$$T_{pc}(R) = 169.349.5\gamma_g - 74\gamma_g^2 \tag{2.2}$$

Sutton used the 264 measured z factor data points of Dranchuk and Abou-Kassem (1975) and used Wichert and Aziz (1971) for adjustment of the non-hydrocarbons effect.

However, if critical properties can be modeled in compositional form, such models as the models of Whitson (1982) and Nikitin and Popov (2016) can be used. It should be mentioned that using compositional modeling in nodal analysis makes the calculation very complex and tedious.

Another property is viscosity which its estimation is the subject of the study of Alomair et al. (2014), (Hemmati-Sarapardeh et al. 2014; Ghorbani et al. 2016). There are different correlations for estimating the oil viscosity. At pressure above the bubble point, viscosity changes linearly with pressure. Thus its estimation below bubble point has a greater importance and most correlations estimated the viscosity at that pressure (Naji 2013). Table 2.1 shows some of the most common correlations for estimating the oil viscosity.

Another parameter is the gas compressibility factor, which is necessary in nodal analysis calculation (Kamari et al. 2016; Zheng et al. 2016). Table 2.2 shows some correlations for estimating this property. Table 2.2 shows some of the most common methods for estimating the gas compressibility factor.

And finally for the solution gas oil ratio references such as (Standing 1947; Gharbi and Elsharkawy 1997; Tohidi-Hosseini et al. 2016) can be studied.

Table 2.1 Correlations for estimating the oil viscosity

Correlation	Equations
Beal (1946)	$\mu_o = \mu_{ob} + [0.001(P - P_b)]\left(0.024\mu_{ob}^{1.6} + 0.038\mu_{ob}^{0.56}\right)$
Kouzel (1965)	$\mu_o = \mu_{ob}e^{\alpha(P-P_b)}$
	$\alpha = 5.50318 * 10^{-5} + 3.77163 * 10^{-5}\mu_{ob}^{0.278}$
Al-Khafaji et al. (1987)	$\mu_o = \mu_{ob} + 10^{[X + 1.11\log(0.07031(P - P_b))]}$
	$X = -0.3806 - 0.1845\gamma_{API} + 0.004034\gamma_{API}^2$
	$\qquad -3.716 * 10^{-5}\gamma_{API}^3$
Khan et al. (1987)	$\mu_o = \mu_{ob}e^{9.6*10^{-5}(P-P_b)}$
Petrosky (1990)	$\mu_o = \mu_{ob} + 1.3449 * 10^{-3}(P - P_b) * 10^{X_2}$
	$X_2 = -1.0146 + 1.3322X_1 - 0.4876X_1^2 - 1.15036X_1^3$
	$X_1 = \log(\mu_{ob})$
Abdul-Majeed et al. (1990)	$\mu_o = \mu_{ob} + 1000 * 10^{[X - 5.2106 + 1.11\log(6.894757(P - P_b))]}$
	$X = 1.9311 - 0.89941(\ln R_{sb}) - 0.001194\gamma_{API}^2$
	$\qquad + 0.0092545\gamma_{API}(\ln(R_{sb}))$
Kartoatmodjo and Schmidt (1991)	$\mu_o = 1.00081\mu_{ob} + 1.127 * 10^{-3}(P - P_b)$
	$\qquad \left(-6.517 * 10^{-3}\mu_{ob}^{1.8148} + 0.038 * \mu_{ob}^{1.59}\right)$
Almehaideb (1997)	$\mu_o = \mu_{ob}\left(\frac{P}{P_b}\right)^{0.134819 + 1.94345*10^{-4}R_{sb} - 1.93106*10^{-9}R_{sb}^2}$
Dindoruk and Christman (2001)	$\mu_o = \mu_{ob} + a_6(P - P_b) * 10^X$
	$X = a_1 + a_2\log(\mu_{ob}) + a_3\log(R_{sb})$
	$\qquad + u_4\mu_{ob}\log(R_{sb}) + a_5(P - P_b)$
	$a_1 = 0.776644$ $\qquad\qquad$ $a_4 = 0.009148$
	$a_2 = 0.987658$ $\qquad\qquad$ $a_5 = -0.000019111$
	$a_3 = -0.19056$ $\qquad\qquad$ $a_6 = 0.000063340$

(continued)

Table 2.1 (continued)

Correlation	Equations
Bergman and Sutton (2009)	$\mu_o = \mu_{ob} e^{\alpha(P-P_b)^\beta}$
	$\alpha = 6.6598 * 10^{-6} \ln(\mu_{ob})^2$ $-1.4821 * 10^{-5} \ln(\mu_{ob}) + 2.27877 * 10^{-4}$
	$\beta = 2.24623 * 10^{-2} \ln(\mu_{ob}) + 0.873204$
Hemmati-Sarapardeh et al. (2013)	$\mu_{od} = \frac{A}{B} e^{\frac{C}{D}}$
	$A = T^2 + a_1 T + a_2$
	$B = API^2 + a_3 * API + a_4$
	$C = a_5 T + a_6$
	$D = a_7 * API + a_8$
	$a_1 = -160.0514$ $\qquad$ $a_2 = 12488.07$
	$a_3 = 3482.605$ $\qquad$ $a_4 = -43254.99$
	$a_5 = -0.004525228$ $\qquad$ $a_6 = 1.329148$
	$a_7 = 0.004335506$ $\qquad$ $a_8 = 0.08006255$
Abooali and Khamehchi (2014)	$\mu_g = 0.007393 + 0.2738481408 \left(\dfrac{\rho_g}{T_{pr}}\right)^2$ $+ 0.594577152 \left(\dfrac{\rho_g^2 P_{pr}}{62.4\rho_g + P_{pr}}\right)$ $-1.5620581417 - 10^{-3} \left(\rho_g^3\right)\left(M_a + 62.4\rho_g\right) + 9.59 * 10^{-5}\left(M_a \times T_{pr}^2\right)$
Ghorbani et al. (2016)	$y = Aa$
	$a = (a_0, a_1, a_2, a_3, a_4, a_5)$
	This method is an iterative one and the value of parameters for different cases is listed in the paper

In addition, a relation is required for estimating the temperature profile in the tubing. There are some studies that have focused in estimating the temperature profile in the well such as (Hasan and Kabir 1991; Cazarez-Candia and Vásquez-Cruz 2005; Yoshioka et al. 2005; Espinosa-Paredes et al. 2009; Mahdiani and Khamehchi 2016). In addition, as a correlation for the multiphase flow in the tubing is necessary, there are some researches that are focused specifically on gas lift cases such as (Poettman and Carpenter 1952; Tek 1961; Fancher Jr. and Brown 1963; Weisman and Kang 1981; Kolev and Kolev 2005; Guet and Ooms 2006).

The nodal analysis calculations continue until the pressure at the end of the section converges. Afterward, this process continues for the next section and at the end the bottom hole pressure for different production rates is calculated. Finally, using that and the reservoir delivery models, the bottom hole pressure and production rate is calculated (Fattah et al. 2014). It is clear that in gas lift calculation the effect of gas lift is considered by a different gas liquid ratio higher and lower

Table 2.2 Some correlations for estimating the gas compressibility factor

Correlation	Equations
Beggs and Brill (1973)	$z = A + \frac{(1-A)}{e^B} + C\left(\frac{P}{P_{pc}}\right)^D$
	$A = 1.39\left(\frac{T}{T_{pc}} - 0.92\right)^{0.5} - 0.36\frac{T}{T_{pc}} - 0.1$
	$B = \left(0.62 - 0.23\frac{T}{T_{pc}}\right) * \left(\frac{P}{P_{pc}}\right)$
	$+ \left(\frac{0.066}{\frac{T}{T_{pc}} - 0.86} - 0.37\right)\left(\frac{P}{P_{pc}}\right)^2 + \frac{0.32\left(\frac{P}{P_{pc}}\right)^6}{10^E}$
	$C = 0.132 - 0.32\log\left(\frac{T}{T_{pc}}\right)$
	$D = 10^F$
	$E = 9 * \left(\frac{T}{T_{pc}} - 1\right)$
	$F = 0.3106 - 0.49\left(\frac{T}{T_{pc}}\right) + 0.1824\left(\frac{T}{T_{pc}}\right)^2$
Kumar (2004)	$Z = A + BP_{pr} + (1-A)\exp(-C) - D\left(\frac{P_{pr}}{10}\right)^4$
	$A = -0.101 - 0.36T_{pr} + 1.3868\sqrt{T_{pr} - 0.919}$
	$B = 0.021 + \frac{0.04275}{T_{pr} - 0.65}$
	$C = P_{pr}\left(E + FP_{pr} + GP_{pr}^4\right)$
	$D = 0.122\exp\left(-11.3\left(T_{pr} - 1\right)\right)$
	$E = 0.622 - 0.224T_{pr}$
	$F = \frac{0.0657}{T_{pr} - 0.85} - 0.037$
	$G = 0.32\exp\left(-19.53\left(T_{pr} - 1\right)\right)$
Heidaryan et al. (2010)	$z = \dfrac{A_1 + A_2\ln(p_{pr}) + A_3\left(\ln p_{pr}\right)^2 + A_4\left(\ln p_{pr}\right)^3 + \frac{A_5}{T_{pr}} + \frac{A_6}{T_{pr}^2}}{1 + A_7\ln(p_{pr}) + A_8\left(\ln p_{pr}\right)^2 + \frac{A_9}{T_{pr}} + \frac{A_{10}}{T_{pr}^2}}$
	A_1 1.11532372699824
	A_2 -0.07903952088760
	A_3 0.01588138045027
	A_4 -0.00886134496010
	A_5 -2.16190792611599
	A_6 1.15753118672070
	A_7 -0.05367780720737
	A_8 0.014655569989618
	A_9 -1.80997374923296
	A_{10} 0.95486038773032
Azizi et al. (2010)	$Z = A + \frac{B+C}{D+E}$
	$A = aT_{pr}^{2.16} + bP_{pr}^{1.028} + CP_{pr}^{1.58}T_{pr}^{-2.1} + d\ln\left(T_{pr}\right)^{-0.5}$
	$B = e + fT_{pr}^{2.4} + gP_{pr}^{1.56} + hP_{pr}^{0.124}T_{pr}^{3.033}$

(continued)

Table 2.2 (continued)

Correlation	Equations			
	$C = i \ln(T_{pr})^{-1.28} + j\ln(T_{pr})^{1.37} + k\ln(P_{pr})$ $+ l\ln(P_{pr})^2 + m\ln(P_{pr})\ln(T_{pr})$			
	$D = 1 + nT_{pr}^{5.5} + oP_{pr}^{0.68}T_{pr}^{0.33}$			
	$E = p\ln(T_{pr})^{1.18} + q\ln(T_{pr})^{2.1} + r\ln(P_{pr})$ $+ s\ln(P_{pr})^2 + t\ln(P_{pr})\ln(T_{pr})$			
	a	0.0373142485385592	k	−24449114791.1531
	b	−0.0140807151485369	l	19357955749.3274
	c	0.0163263245387186	m	−126354717916.607
	d	−0.0307776478819813	n	623705678.385784
	e	13843575480.943800	o	17997651104.3330
	f	−16799138540.763700	p	151211393445.064
	g	1624178942.6497600	q	139474437997.172
	h	13702270281.086900	r	−242330112984.0950
	i	−41645509.896474600	s	18938047327.5205
	j	237249967625.01300	t	−141401620722689
Sanjari and Lay (2012)	$z = 1 + A_1 P_{pr} + A_2 P_{pr}^2 + \frac{A_3 P_{pr}^{A_4}}{T_{pr}^{A_5}} + \frac{A_6 P_{pr}^{(A_4+1)}}{T_{pr}^{A_7}} + \frac{A_8 P_{pr}^{A_4+2}}{T_{pr}^{A_7+1}}$			
	A_1	0.007698		
	A_2	0.003839		
	A_3	−0.467212		
	A_4	1.018801		
	A_5	3.8057233		
	A_6	−0.087361		
	A_7	7.138305		
	A_8	0.083440		
Fatoorehchi et al. (2014)	$z = 1 + \left(A_4 T_{pr} - A_2 - \frac{A_6}{T_{pr}^2}\right)\left(\frac{P_{pr}}{zT_{pr}^2}\right) + (A_3 T_{pr} - A_1)\left(\frac{P_{pr}^2}{z^2 T_{pr}^3}\right)$ $+ \frac{A_1 A_5 A_7 P_{pr}^5}{z^5 T_{pr}^6}\left(1 + \frac{A_8 P_{pr}^2}{z^2 T_{pr}^2}\right)\exp\left(-\frac{A_8 P_{pr}^2}{z^2 T_{pr}^2}\right)$			
	Coefficient	0.4 < p_pr < 5.00.4 < p_pr < 5.0	5 ≤ p_pr < 155 ≤ ppr < 15	
	A_1	0.001290236	0.0014507882	
	A_2	0.38193005	0.37922269	
	A_3	0.022199287	0.024181399	
	A_4	0.12215481	0.11812287	
	A_5	−0.015674794	0.037905663	
	A_6	0.027271364	0.19845016	
	A_7	0.023834219	0.048911693	
	A_8	0.43617780	0.0631425417	

(continued)

Table 2.2 (continued)

Correlation	Equations
Kamari et al. (2016)	$Z = 0.2625136 + \dfrac{3.1263651}{T_{pr}} + \dfrac{-3.8916368}{T_{pr}^2}$
	$+ \dfrac{1.0551763}{T_{pr}^3} + 0.56388 \ln P_{pr}$
	$- 0.3372525 \left(\ln\left(P_{pr}\right)\right)^2 + 0.061688 \left(\ln\left(P_{pr}\right)\right)^3$
	$+ \dfrac{-1.3976452 \ln\left(P_{pr}\right)}{T_{pr}} + \dfrac{0.5217521 \ln\left(P_{pr}\right)}{T_{pr}^2} + \dfrac{0.447935 \ln\left(P_{pr}\right)^2}{T_{pr}}$

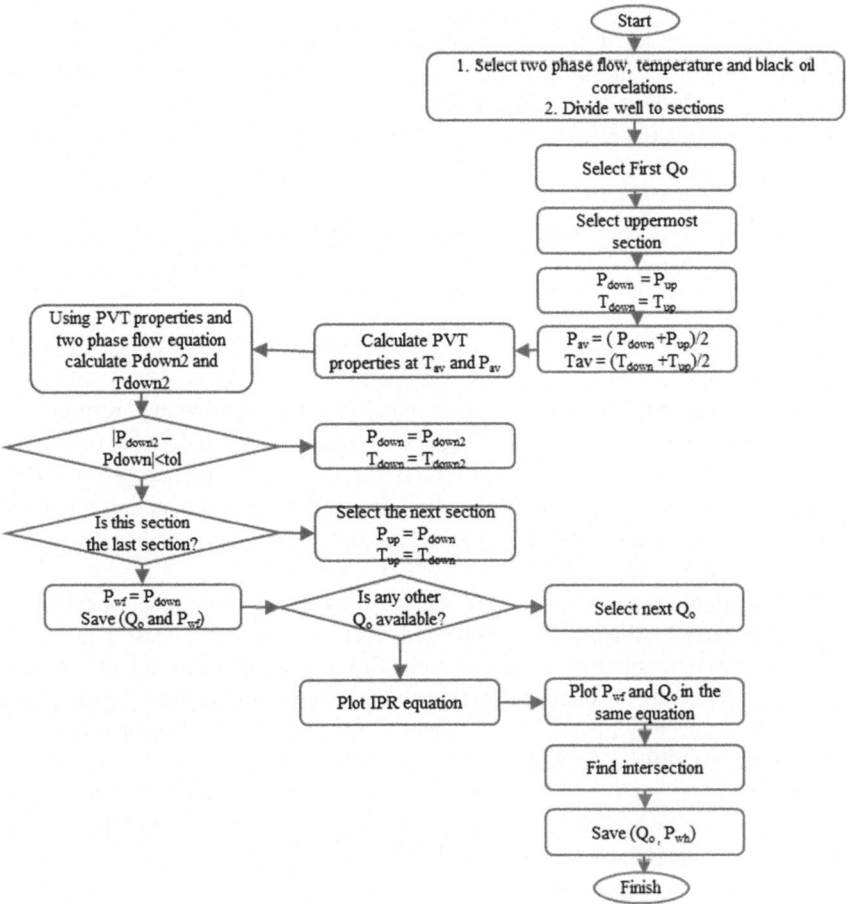

Fig. 2.1 Flowchart of nodal analysis (Mahdiani and Khamehchi 2015)

than the injection point in the multiphase model. A flowchart of nodal analysis is illustrated in Fig. 2.1.

Some studies used this model for developing the fitness function (Kelkar 2008) and the general form of the nodal analysis is as above, but in details it has minor differences in different studies (Hamedi et al. 2011). (Rashidi et al. 2010) used the GOPUGS software as its fitness function. This software is based on nodal analysis. They used its model in an optimization problem, not a gas allocation one but for optimizing single well parameters.

2.2.2 Analytical Models

As mentioned earlier it is very common to use empirical correlations in nodal analysis. However, in some studies an analytical overview can also be used. Imsland et al. (2003) considered a section in the annulus and tubing and used mass balance equations to create an analytic model for that. His basic equations were as below:

$$\dot{x}_1 = -\omega_{gc}(x) + \omega_{iv}(x, u_1) \tag{2.3}$$

$$\dot{x}_2 = -\omega_{iv}(x) - \omega_{pg}(x, u_2) \tag{2.4}$$

$$\dot{x}_3 = \omega_r(x) - \omega_{po}(x, u_2) \tag{2.5}$$

The first two equations show the mass of gas in annulus and tubing, respectively, and the third one is mass of oil in tubing. These equations are in derivative form and are good in control studies. In them, ω_{gc} is the flow of gas through gas injection choke, ω_{pg} and ω_{po} are the gas flow and oil flow through the production choke and $\omega_r(x)$ is the flow of oil from the reservoir. Imsland (2002) also introduced a relation between the flow and pressure of the system.

Here a simple model of this kind will be illustrated. This model is called the third-order model (Shao et al. 2016b) and assumes a constant mass ratio of the oil/gas in different points of the tubing and a fast and homogenous mixing between oil and gas. This model is not able to explain the dynamic of the gas lift systems and is just applicable when the system has gained the steady state. Consider the schematic of the gas lift in Fig. 2.2.

Here the ideal gas is assumed. Then by considering the momentum balance of gas, the pressure at the injection point p_{ai} and the pressure at the top of the annulus, p_{at} can be calculated as below (Imsland 2002):

$$p_{ai} = \frac{m_{ga}g}{A_a} \frac{1}{1 - e^{\frac{gM}{RT_a}H_a}} \tag{2.6}$$

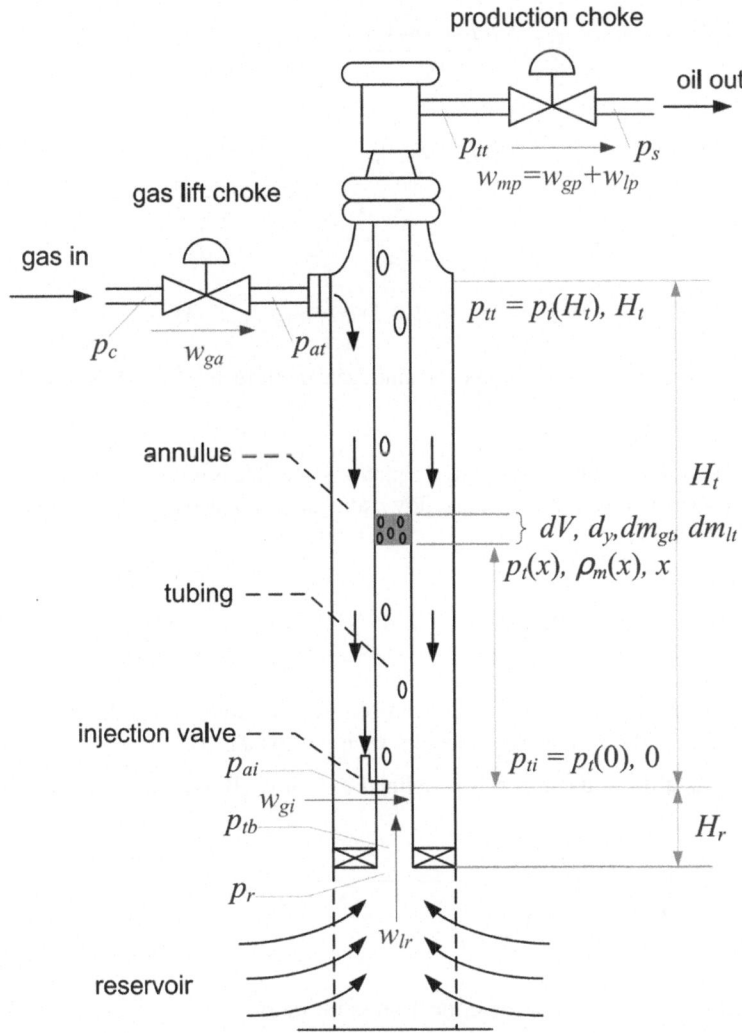

Fig. 2.2 A simple model for gas lift (Shao et al. 2016a)

$$p_{ta} = p_{ai}e^{-\frac{gM}{RT_a}H_a} = \frac{m_{ga}}{A_a}\frac{e^{-\frac{gM}{RT_a}H_a}}{1 - e^{-\frac{gM}{RT_a}H_a}} \tag{2.7}$$

Here the gas oil ratio is defined as $\lambda = m_{gt}/m_{lt}$ and due to the homogenous assumption, λ is constant in all the points of the tubing. Then the pressure at the injection point is as follows:

$$p_{ti} = \frac{(m_{gt} + m_{lt})g}{A_t} \frac{1}{1 - e^{\frac{(m_{gt} + m_{lt})g - aA_tH_t}{A_tb}}} \tag{2.8}$$

$$p_{tt} = \frac{(m_{gt} + m_{lt})g}{A_t} \frac{e^{\frac{(m_{gt} + m_{lt})g - aA_tH_t}{A_tb}}}{1 - e^{\frac{(m_{gt} + m_{lt})g - aA_tH_t}{A_tb}}} \tag{2.9}$$

$$p_{tb} = p_{ti} + \rho g H_r \tag{2.10}$$

$$\text{Where } a = \rho_l g (1 + \lambda) \text{ and } b = \lambda \rho_l T_t / M. \tag{2.11}$$

In addition the density of the oil and gas mixture at the top of tubing is as follows: $\rho_m = \frac{1+\lambda}{\lambda \frac{RT}{p_{tt}M} + \frac{1}{\rho_l}}$

For more details about the above equations see (Hussein et al. 2015).

As shown in Fig. 2.2, the mass flow rates are shown by "w" and can be calculated as below:

$$w_{ga} = 7.309 * 10^{-6} c_v p_c u_{gc} \sqrt{\frac{g\gamma}{R_{mg}T_a} \left(\frac{2}{\gamma+1}\right)^{\frac{\gamma+1}{\gamma-1}} \Psi\left(\frac{p_{at}}{p_c}\right)} \tag{2.12}$$

where c_v is the valve constant, it depends on the size and type of the valve and on this basis it can be found in handbooks such as (Fisher 2005).

$\Psi\left(\frac{p_2}{p_1}\right)$ is a flow function and is defined as below (Boiko and Sayedain 2010):

$$\Psi\left(\frac{p_2}{p_1}\right) = \begin{cases} 1 & \text{if } \frac{p_2}{p_1} < \beta_c \\ \sqrt{\frac{2}{\gamma-1}\left(\frac{\gamma+1}{2}\right)^{\frac{\gamma+1}{\gamma-1}}} \sqrt{\left(\frac{p_2}{p_1}\right)^{\frac{2}{\gamma}} - \left(\frac{p_2}{p_1}\right)^{\frac{\gamma+1}{\gamma}}} & \text{if } \frac{p_2}{p_1} \geq \beta_c \end{cases} \tag{2.13}$$

Where $\beta_c = \left(\frac{2}{\gamma+1}\right)^{\frac{\gamma}{\gamma-1}}$ is the critical pressure ratio.

The gas mass flow from the tubing to annulus can be calculated as below:

$$w_{gi} = A_{iv} c_{iv} p_{ai} \sqrt{\frac{g\gamma}{R_{ng}T_a} \left(\frac{2}{\gamma+1}\right)^{\frac{\gamma+1}{\gamma-1}} \Psi\left(\frac{p_{ti}}{p_{ai}}\right)} \tag{2.14}$$

And the mass flow rate of oil which flows from the reservoir to tubing is:

$$w_{lr} = \rho_l q_{lr} \tag{2.15}$$

In the above formula, q_{lr} is the volumetric flow rate.

$$q_{lr} = N_{lr} A_{lr} c_{lr} \sqrt{\frac{p_r - p_{tb}}{g_l}} \qquad (2.16)$$

In (2.16) the effective area of the orifice is provided by $A_{lr}c_{lr}$, N_{lr} is a constant for unit conversion and $g_l = \frac{\rho_l}{1000}$ is the specific gravity of the oil. The mass flow rate can be calculated as:

$$w_{mp} = \rho_m q_{mp} \qquad (2.17)$$

q_{mp} is the volumetric flow rate of the mixture in the tubing and can be calculated as below:

$$q_{mp} = k c_v u_{pc} \sqrt{\frac{p_{tt} - p_s}{g_m}} \qquad (2.18)$$

In the above, k is obtained from unit conversion and the definition of the c_v.

Finally, the gas mass flow rate and the oil flow rate through the production choke can be given by:

$$w_{gp} = \frac{\lambda}{1+\lambda} w_{mp} \qquad (2.19)$$

$$w_{lp} = \frac{1}{1+\lambda} w_{mp} \qquad (2.20)$$

And finally, the balance equation for the mass flow rates of gas in annulus and tubing and the oil in tubing can be expressed as below:

$$\dot{m}_{ga} = w_{ga} - w_{gi} \qquad (2.21)$$

$$\dot{m}_{gt} = w_{gi} - w_{gp} \qquad (2.22)$$

$$\dot{m}_{lt} = w_{lr} - w_{lp} \qquad (2.23)$$

The above equations can be solved by numerical methods.

2.2.3 Proxy Models

Nodal analysis is much more accurate than proxy models, but the problem is that it is very slow and time consuming. Thus, in optimization problems which need a huge number of calculations of the fitness function, using some faster methods is interesting (Panjalizadeh et al. 2014; Golzari et al. 2015). Proxy models are trained by some previously measured points and based on them the value of some new

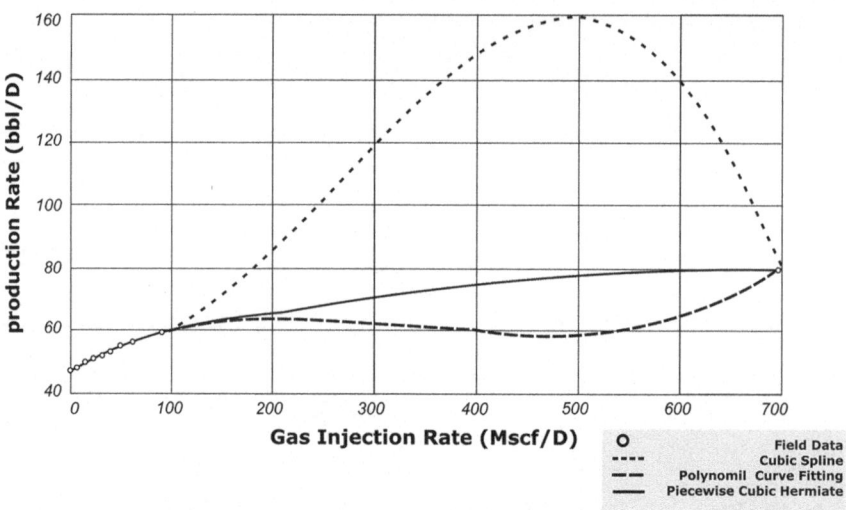

Fig. 2.3 Comparison between some proxy models is the prediction of the oil rate of a typical well based on the injection rate (Ghassemzadeh et al. 2015)

points can be estimated (Khamehchi et al. 2009, 2014). There are different types of proxy models such as cubic spline (Fitra et al. 2015), polynomial curve fitting (Monyei et al. 2014), piecewise cubic hermite (Xu and Zhang 2014), neural networks (Chithra Chakra et al. 2013) genetic programming (Kaydani et al. 2014; Mahdiani and Kooti 2016), support vector machine (Hou et al. 2009), least square support vector machine (Kisi and Parmar 2016), co-active neuro-fuzzy inference system, simulated annealing programming (Mahdiani and Khamehchi 2014; Mahdiani et al. 2015), etc. As an example, (Ebrahimi and Khamehchi 2016) used support vector machines for creating a proxy model to estimate the oil rate of a gas lift well. He reasoned its use due to the privileged generalization performance in regression and classification. He developed a proxy for a particular field using SVM. Then, for the optimal parameter control, the simulator coupled it with the particle swarm optimization algorithm. He used this model in a multi objective problem to maximize oil production and simultaneously minimize the water production. Using different proxy models leads to different accuracies (Fig. 2.3).

2.3   Economic Factors

When just one well is considered, the amount of injected gas has an important effect on net present value. Thus, more oil production does not necessarily mean more profit because of the cost of gas injection such as the compressor (Clegg 1981).

On the other hand, usually a compressor with constant power is installed and all its capacity is used. Thus, its cost is fixed. Altering the share of different wells would not change other capital costs, and thus the net present value is proportional to the production rate (Takács 2005). So in many problems, the cumulative production rate is considered as the value of fitness function and the optimizer tries to maximize it. However, some other parameters such as time are still important in the net present value. The back payment is very dominant in calculations and as the pressure of the reservoir decreases, the back payment at different times is not the same. It is clear that for considering this there is a need for the integrated model (Ghassemzadeh and Pourafshary 2015; Khishvand and Khamehchi 2012).

2.4   Integrated Model

In previous parts the considered model was just a model for well performance. For some purposes such as some economical surveys, the well model alone is not sufficient. Here the integrated model is necessary. The integrated model considers the reservoir, well and surface facilities together and their effects on each other are thought-out (Gutierrez et al. 2007).

Some integrated models are made by just coupling the reservoir and well model. They calculate inflow performance relation (IPR) and tubing performance relation (TPR) and find their intersection (Rasouli et al. 2011). In some others, they may consider the surface pipelines and the pressure loss (El-Massry and Price 1995) (Fig. 2.4).

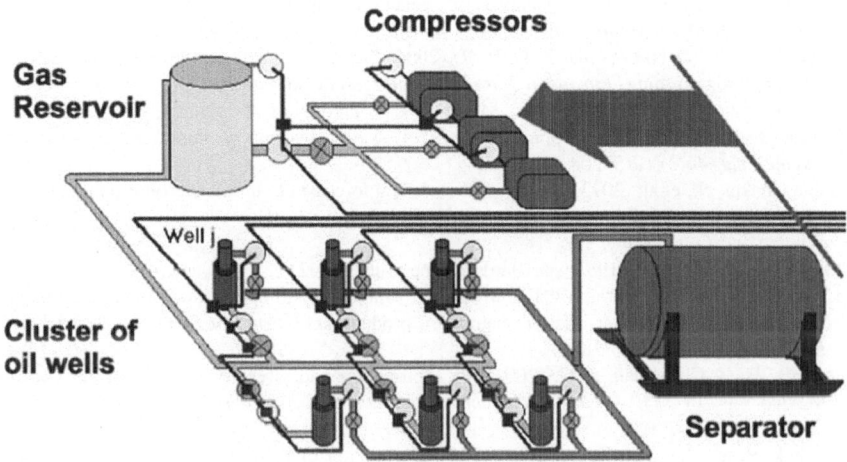

Fig. 2.4 A schematic of the components in an integrated model (Camponogara and Nakashima 2006)

References

Abdul-Majeed, G. H., Clark, K. K., & Salman, N. H. (1990). New correlation for estimating the viscosity of undersaturated crude oils. *Journal of Canadian Petroleum Technology, 29*(03).

Abooali, D., & Khamehchi, E. (2014). Estimation of dynamic viscosity of natural gas based on genetic programming methodology. *Journal of Natural Gas Science and Engineering, 21*, 1025–1031.

Al-Khafaji, A. H., Abdul-Majeed, G. H., & Hassoon, S. F. (1987). Viscosity correlation for dead, live and undersaturated crude oils. *Journal of Pets Research, 6*(2), 1–16.

Almehaideb, R. A. (1997). Improved PVT correlations for UAE crude oils. In *Middle East Oil Show and Conference*. Society of Petroleum Engineers.

Alomair, O., Elsharkawy, A., & Alkandari, H. (2014). A viscosity prediction model for Kuwaiti heavy crude oils at elevated temperatures. *Journal of Petroleum Science and Engineering, 120*, 102–110.

Asadisaghandi, J., & Tahmasebi, P. (2011). Comparative evaluation of back-propagation neural network learning algorithms and empirical correlations for prediction of oil PVT properties in Iran oilfields. *Journal of Petroleum Science and Engineering, 78*(2), 464–475.

Asoodeh, M., & Bagheripour, P. (2012). Estimation of bubble point pressure from PVT data using a power-law committee with intelligent systems. *Journal of Petroleum Science and Engineering, 90–91*, 1–11.

Avaullee, L., et al. (1997). Thermodynamic modeling for petroleum fluids I. Equation of state and group contribution for the estimation of thermodynamic parameters of heavy hydrocarbons. *Fluid Phase Equilibria, 139*(1), 155–170.

Azizi, N., Behbahani, R., & Isazadeh, M. A. (2010). An efficient correlation for calculating compressibility factor of natural gases. *Journal of Natural Gas Chemistry, 19*(6), 642–645.

Beal, C. (1946). The viscosity of air, water, natural gas, crude oil and its associated gases at oil field temperatures and pressures. *Transactions of the AIME, 165*(01), 94–115.

Beggs, D. H., & Brill, J. P. (1973). A study of two-phase flow in inclined pipes. *Journal of Petroleum Technology, 25*(05), 607–617.

Bergman, D. F., & Sutton, R. P. (2009). A consistent and accurate dead-oil-viscosity method. *SPE Reservoir Evaluation & Engineering, 12*(06), 815–840.

Boiko, I., & Sayedain, S. (2010). Analysis of dynamic nonlinearity of flow control loop through modified relay test probing. *International Journal of Control, 83*(12), 2580–2587.

Camponogara, E., & Nakashima, P. H. R. R. (2006). Solving a gas-lift optimization problem by dynamic programming. *European Journal of Operational Research, 174*(2), 1220–1246.

Cazarez-Candia, O., & Vásquez-Cruz, M. A. (2005). Prediction of pressure, temperature, and velocity distribution of two-phase flow in oil wells. *Journal of Petroleum Science and Engineering, 46*(3), 195–208.

Chithra Chakra, N., et al. (2013). An innovative neural forecast of cumulative oil production from a petroleum reservoir employing higher-order neural networks (HONNs). *Journal of Petroleum Science and Engineering, 106*, 18–33.

Clegg, J. D. (1981). kanu_1982_spe10865.pdf. pp. 1887–1892.

de Souza, J. N. M. N. M., et al. (2010). Modeling, simulation and optimization of continuous gas lift systems for deepwater offshore petroleum production. *Journal of Petroleum Science and Engineering, 72*(3), 277–289.

Dindoruk, B., & Christman, P. G. (2001). PVT properties and viscosity correlations for Gulf of Mexico oils. In *SPE Annual Technical Conference and Exhibition*. Society of Petroleum Engineers.

Dranchuk, P. M., & Abou-Kassem, H. (1975). Calculation of Z factors for natural gases using equations of state. *Journal of Canadian Petroleum Technology, 14*(03).

Ebrahimi, A., & Khamehchi, E. (2016). Developing a novel workflow for natural gas lift optimization using advanced support vector machine. *Journal of Natural Gas Science and Engineering, 28*, 626–638.

El-Massry, Y. A.-W., & Price, A. D. (1995). Development of a network and gas lift allocation model for production optimization in the ras budran field. In *Proceedings of Middle East Oil Show*. Society of Petroleum Engineers.

Espinosa-Paredes, G., et al. (2009). Application of a proportional-integral control for the estimation of static formation temperatures in oil wells. *Marine and Petroleum Geology, 26*(2), 259–268.

Fancher, G. H., Jr., & Brown, K. E. (1963). Prediction of pressure gradients for multiphase flow in tubing. *Society of Petroleum Engineers Journal, 3*(01), 59–69.

Fatoorehchi, H., Abolghasemi, H., & Rach, R. (2014). An accurate explicit form of the Hankinson–Thomas–Phillips correlation for prediction of the natural gas compressibility factor. *Journal of Petroleum Science and Engineering, 117*, 46–53.

Fattah, K. A., et al. (2014). New inflow performance relationship for solution-gas drive oil reservoirs. *Journal of Petroleum Science and Engineering, 122*, 280–289.

Fisher. (2005). Control valve handbook Fourth.

Fitra, U. R., Rahmawati, S. D., Sukarno, P., & Soewono, E. (2015). Optimization of gas lift allocation in multi-well system, a simple numerical approach. In *Proceedings, Indonesian Petroleum Association*.

Gharbi, R. B., & Elsharkawy, A. M. (1997). Neural network model for estimating the PVT properties of Middle East crude oils. In *Middle East Oil Show and Conference*. Society of Petroleum Engineers.

Ghassemzadeh, S., & Pourafshary, P. (2015). Development of an intelligent economic model to optimize the initiation time of gas lift operation. *Journal of Petroleum Exploration and Production Technology, 5*(3), 315–320.

Ghassemzadeh, S., et al. (2015). Optimization of gas lift allocation for improved oil production under facilities constraints. *Geosystem Engineering, 5*(3), 39–47.

Ghorbani, B., et al. (2016). A novel multi-hybrid model for estimating optimal viscosity correlations of Iranian crude oil. *Journal of Petroleum Science and Engineering, 142*, 68–76.

Golzari, A., Haghighat Sefat, M., & Jamshidi, S. (2015). Development of an adaptive surrogate model for production optimization. *Journal of Petroleum Science and Engineering, 133*, 677–688.

Guet, S., & Ooms, G. (2006). Fluid mechanical aspects of the gas-lift technique. *Annual Review of Fluid Mechanics, 38*, 225–249.

Gutierrez, F. A., et al. (2007). A new approach to gas lift optimization using an integrated asset model. In *International Petroleum Technology Conference*. International Petroleum Technology Conference.

Hamedi, H., Rashidi, F., & Khamehchi, E. (2011). Numerical prediction of temperature profile during gas lifting. *Petroleum Science and Technology, 29*, 1305–1316.

Hasan, A. R., & Kabir, C. S. (1991). Heat transfer during two-Phase flow in Wellbores; Part I–formation temperature. In *SPE Annual Technical Conference and Exhibition*. Society of Petroleum Engineers.

Heidaryan, E., Salarabadi, A., & Moghadasi, J. (2010). A novel correlation approach for prediction of natural gas compressibility factor. *Journal of Natural Gas Chemistry, 19*(2), 189–192.

Hemmati-Sarapardeh, A., et al. (2013). Toward reservoir oil viscosity correlation. *Chemical Engineering Science, 90*, 53–68.

Hemmati-Sarapardeh, A., et al. (2014). Reservoir oil viscosity determination using a rigorous approach. *Fuel, 116*, 39–48.

Hou, J., et al. (2009). Integrating genetic algorithm and support vector machine for polymer flooding production performance prediction. *Journal of Petroleum Science and Engineering, 68*(1–2), 29–39.

Hussein, H., Al-Durra, A., & Boiko, I. (2015). Design of gain scheduling control strategy for artificial gas lift in oil production through modified relay feedback test. *Journal of the Franklin Institute, 352*(11), 5122–5144.

Imsland, L. (2002). Topics in nonlinear control: Output feedback stabilization and control of positive systems.

Imsland, L., Foss, B. A., & Eikrem, G. O. (2003). State feedback control of a class of positive systems: Application to gas-lift stabilization. *European Control Conference, ECC, 2003*, 2499–2504.

Joback, K. G., & Reid, R. C. (1987). Estimation of pure-component properties from group-contributions. *Chemical Engineering Communications, 57*(1–6), 233–243.

Kamari, A., et al. (2016). A corresponding states-based method for the estimation of natural gas compressibility factors. *Journal of Molecular Liquids, 216*, 25–34.

Kartoatmodjo, T., & Schmidt, Z. (1991). New correlations for crude oil physical properties. *SPE Paper, 23556*, 1–11.

Kaydani, H., Mohebbi, A., & Eftekhari, M. (2014). Permeability estimation in heterogeneous oil reservoirs by multi-gene genetic programming algorithm. *Journal of Petroleum Science and Engineering, 123*, 201–206.

Kelkar, M. (2008). *Natural gas production engineering*. PennWell Books.

Khamehchi, E., et al. (2009). Intelligent system for continuous gas lift operation and design with unlimited gas supply. *Journal of Applied Sciences, 9*(10), 1889–1897.

Khamehchi, E., Abdolhosseini, H., & Abbaspour, R. (2014). Prediction of maximum oil production by gas lift in an Iranian field using auto-designed neural network. *Academic Research Online Publisher, 2*(2), 138–150.

Khan, S. A., et al. (1987). Viscosity correlations for Saudi Arabian crude oils. In *Middle East Oil Show*. Society of Petroleum Engineers.

Khishvand, M., & Khamehchi, E. (2012). Nonlinear risk optimization approach to gas lift allocation optimization. *Industrial and Engineering Chemistry Research, 51*(6), 2637–2643.

Kisi, O., & Parmar, K. S. (2016). Application of least square support vector machine and multivariate adaptive regression spline models in long term prediction of river water pollution. *Journal of Hydrology, 534*, 104–112.

Klincewicz, K. M., & Reid, R. C. (1984). Estimation of critical properties with group contribution methods. *AIChE Journal, 30*(1), 137–142.

Kolev, N. I., & Kolev, N. I. (2005). *Multiphase flow dynamics 2*. Springer.

Kouzel, B. (1965, March). How pressure affects liquid viscosity. *Hydrocarb Process, 120*.

Kumar, N. (2004). Compressibility factors for natural and sour reservoir gases by correlations and cubic equations of state.

Li, J., Xia, L., & Xiang, S. (2016). A new method based on elements and chemical bonds for organic compounds critical properties estimation. *Fluid Phase Equilibria, 417*, 1–6.

Liu, T., et al. (2015). Modelling and extremum seeking control of gas lifted oil wells. *Journal of Petroleum Science and Engineering, 26*(3–4), 21–26.

Mahdiani, M. R., et al. (2015). A new proxy model, based on meta heuristic algorithms for estimating gas compressor torque. In *11th International Industrial Engineering Conference*, Tehran.

Mahdiani, M. R., & Khamehchi, E. (2014). A new method for building proxy models using simulated annealing. *Middle East Journal of Scientific Research, 22*(3), 324–328.

Mahdiani, M. R., & Khamehchi, E. (2015). Stabilizing gas lift optimization with different amounts of available lift gas. *Journal of Natural Gas Science and Engineering, 26*, 18–27.

Mahdiani, M. R., & Khamehchi, E. (2016). A novel model for predicting the temperature profile in gas lift wells. *Petroleum*.

Mahdiani, M. R., & Kooti, G. (2016). The most accurate heuristic-based algorithms for estimating the oil formation volume factor. *Petroleum, 2*(1), 40–48.

Mohsen-Nia, M. (2014). A modified MMM EOS for high-pressure PVT calculations of heavy hydrocarbons. *Journal of Petroleum Science and Engineering, 113*, 97–103.

Monyei, C. G., Adewumi, A. O., & Obolo, M. O. (2014). *Oil well characterization and artificial gas lift optimization using neural networks combined with genetic algorithm*. Discrete Dynamics in Nature and Society.

Naji, H. S. (2013). The oil viscosity correlations: A simulation approach. *Petroleum Science and Technology, 31*(13), 1406–1412.

Nikitin, E. D., & Popov, A. P. (2016). Vapor–liquid critical properties of components of biodiesel. 2. Ethyl esters of n-alkanoic acids. *Fuel, 166*, 502–508.

Panjalizadeh, H., Alizadeh, N., & Mashhadi, H. (2014). A workflow for risk analysis and optimization of steam flooding scenario using static and dynamic proxy models. *Journal of Petroleum Science and Engineering, 121*, 78–86.

Petrosky, G. E. (1990). PVT correlations for gulf of mexico crude oils.

Poettman, F. H., & Carpenter, P. G. (1952). The multiphase flow of gas, oil, and water through vertical flow strings with application to the design of gas-lift installations. In *Drilling and Production Practice*. American Petroleum Institute.

Rashidi, F., Khamehchi, E., & Rasouli, H. (2010). Oil field optimization based on gas lift optimization. In *20th European Symposium on Computer Aided Process Engineering— ESCAPE20*.

Rasouli, H., Rashidi, F., & Khamehchi, E. (2011). Optimization of an integrated model to enhance oil production based on gas lift optimization under limited gas supply. *Oil Gas European Magazine, 37*(4), 199–202.

Sanjari, E., & Lay, E. N. (2012). An accurate empirical correlation for predicting natural gas compressibility factors. *Journal of Natural Gas Chemistry, 21*(2), 184–188.

Shao, W., Boiko, I., & Al-Durra, A. (2016a). Control-oriented modeling of gas-lift system and analysis of casing-heading instability. *Journal of Natural Gas Science and Engineering, 29*, 365–381.

Shao, W., Boiko, I., & Al-Durra, A. (2016b). Plastic bag model of the artificial gas lift system for slug flow analysis. *Journal of Natural Gas Science and Engineering, 33*, 573–586.

Standing, M. B. (1947). A pressure-volume-temperature correlation for mixtures of California oils and gases. In *Drilling and Production Practice*. American Petroleum Institute.

Sutton, R. P. (1985). Compressibility factors for high-molecular-weight reservoir gases. In *SPE Annual Technical Conference and Exhibition*. Society of Petroleum Engineers.

Takács, G. (2005). *Gas Lift Manual*. PennWell.

Tek, M. R. (1961). Multiphase flow of water, oil and natural gas through vertical flow strings. *Journal of Petroleum Technology, 13*(10), 1–29.

Tohidi-Hosseini, S.-M., et al. (2016). Toward prediction of petroleum reservoir fluid properties: A rigorous model for estimation of solution gas-oil ratio. *Journal of Natural Gas Science and Engineering*.

Weisman, J., & Kang, S. Y. (1981). Flow pattern transitions in vertical and upwardly inclined lines. *International Journal of Multiphase Flow, 7*(3), 271–291.

Whitson, C. H. (1982). *Effect of physical properties estimation on equation-of-state predictions*. Rogaland Regional College.

Whitson, C. H. (1984). Critical properties estimation from an equation of state. In *SPE Enhanced Oil Recovery Symposium*. Society of Petroleum Engineers.

Wichert, E., & Aziz, K. (1971). Compressibility factor of sour natural gases. *The Canadian Journal of Chemical Engineering, 49*(2), 267–273.

Xu, G., & Zhang, Z. (2014). Simultaneous approximation of sobolev classes by piecewise cubic hermite interpolation. Numerical Mathematics: Theory, Methods and Applications.

Yang, T., et al. (2016). A novel denitration cost optimization system for power unit boilers. *Applied Thermal Engineering, 96*, 400–410.

Yoshioka, K., et al. (2005). A comprehensive model of temperature behavior in a horizontal well. In *SPE Annual Technical Conference and Exhibition*. Society of Petroleum Engineers.

Zheng, J., et al. (2016). Standardized equation for hydrogen gas compressibility factor for fuel consumption applications. *International Journal of Hydrogen Energy, 41*(15), 6610–6617.

Chapter 3
Constraint Optimization

Abstract In gas allocation optimization problems, some amount of gas is allocated between the wells in a way that the total revenue or total production is maximized. In reality, this problem can have some limitations which make the problem a constraint optimization one. One of the most common constraints in this kind of problem is the maximum available lift gas. In fact, the amount of gas in a gas allocation field is rarely unlimited. Thus, the total injection rate of the found solution should not exceed the maximum available lift gas. Another limitation is the capacity of the facilities. Each well has a line pipe with a limited capacity for fluid transportation, and the capacity of the separator is limited and the water treatment unit cannot handle an unlimited amount of produced water. Thus, the solution should consider all the facility limitations. The injection rate of the gas based on the well and reservoir properties should be in a specific range to escape unstable flow in the tubing or instability phenomenon. Instability can cause the production reduction and damages to surface and downhole facilities. Preventing the unstable flow can be another constraint for the problem. In this chapter, all these constraints and their different aspects will be discussed and then the methods to deal with them will be surveyed. In the remainder of this chapter, the basis of the intelligent wells for online control of the constraints will be explained.

Keywords Facility limitation · Stability phenomenon · Gas lift control · Constraint optimization

3.1 Introduction

In an optimization problem, it is very usual to have a constraint for the solution. This means that the desired solution should satisfy some condition, otherwise no matter what its fitness is, it cannot be accepted (Hou et al. 2016; Chithra Chakra et al. 2015).

© The Author(s) 2017
E. Khamehchi and M.R. Mahdiani, *Gas Allocation Optimization Methods in Artificial Gas Lift*, SpringerBriefs in Petroleum Geoscience & Engineering, DOI 10.1007/978-3-319-51451-2_3

3.2   No Limitation

If in a gas allocation there is no limitation, the allocation means nothing. In fact, in this situation, gas lift optimization means to optimize well parameters in design stages for a single well (Takács 2005). In a single well, by increasing the injection rate the production increases but if it increases too high, production decreases because of some other interfering parameters such as friction (Hussein et al. 2015). Thus, for a specific injection rate the production rate is maximum. The amount of injection rate and its corresponding production depends on well properties, some of which are illustrated in Fig. 3.1.

Khamehchi et al. (2014) and Ranjan et al. (2015) used an artificial neural network for predicting the optimum injection rate. They used the data of production flow and well test and created their model based on that. Their model's input were well parameters and the output was the optimum injection rate and its corresponding production rate. It is clear that for this type of optimization there is a model in reality, one which has been created using optimum points and can predict the optimum point of the new cases.

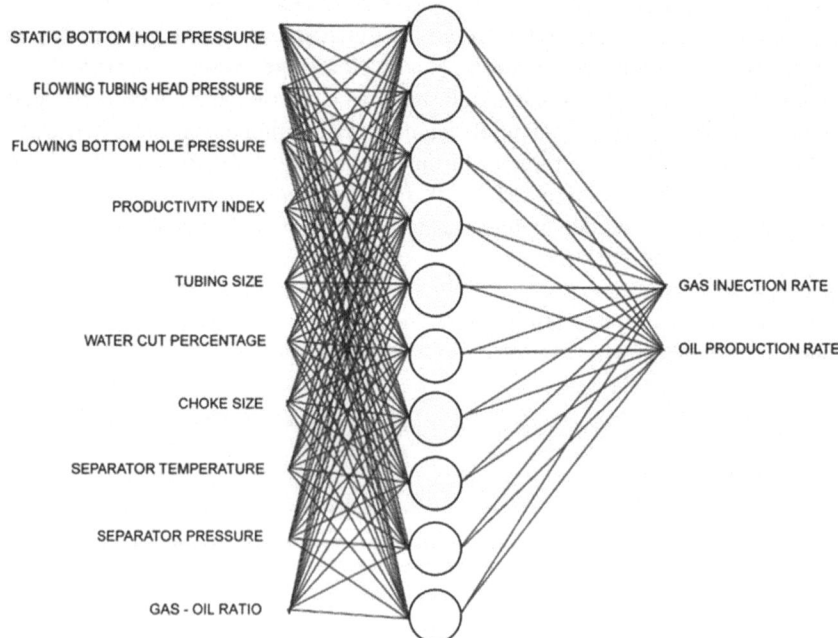

Fig. 3.1 A schematic of a single gas lift well optimization (Ranjan et al. 2015)

3.3 Maximum Amount of Available Lift Gas

In most problems the lift gas is limited. In these problems, the optimizer deals with a constraint optimization. Here, there is a limited amount of lift gas that should be allocated between different wells in a way that the total amount of produced oil is maximized (Takács 2005). This constraint is expressed as (3.1):

$$\sum_{i=1}^{n} q_{gi} \leq Q_{gtotal} \tag{3.1}$$

Ghaedi et al. (2013) considered the limitation in a gas allocation problem. He proposed different cases with different amounts of available lift gas and then optimized them using a hybrid genetic algorithm.

3.4 Injection and Production Limitations

One of the limitations related to production is the water cut of each well (Ebrahimi and Khamehchi 2016). In fact, the water cut should not exceed a maximum. There are some other limitations related to injection and production. For example, in a gas allocation optimization problem, Tapabrata (Ray and Sarker 2007) assumed a maximum limit for the injection rate of each well. Injection pressure depends on the used compressor and due to the difficulty of changing the compressor, in some problems it is considered as a constraint for the optimizer (Rasouli et al. 2011). Some other limitations are such that production should be higher than a minimum (for example to escape liquid loading (Aliev and Ismailov 2015)). In addition, the capacity of facilities in the gas lift system can be a constraint for the problem. This limitation causes the production reduction of other wells.

3.5 Stability

One of the problems in the gas lift process is the instability phenomenon. In this process, the pressure of gas in the wellbore is not high enough to make the wellbore valve continuously open. Thus, it periodically opens and closes and causes high vibration and makes periodical cession to the production (Fairuzov et al. 2004). Figure 3.2 shows the variation of flow rate in a typical unstable flow. As this figure shows, the choke size is one of the parameters that is effective in gas lift flow rate periods.

The pressure and thus rate of injection should be higher than a minimum to escape this phenomenon (Asheim 1988). There are different methods to predict the instability of the flow. One of them which is the most common one is the work of

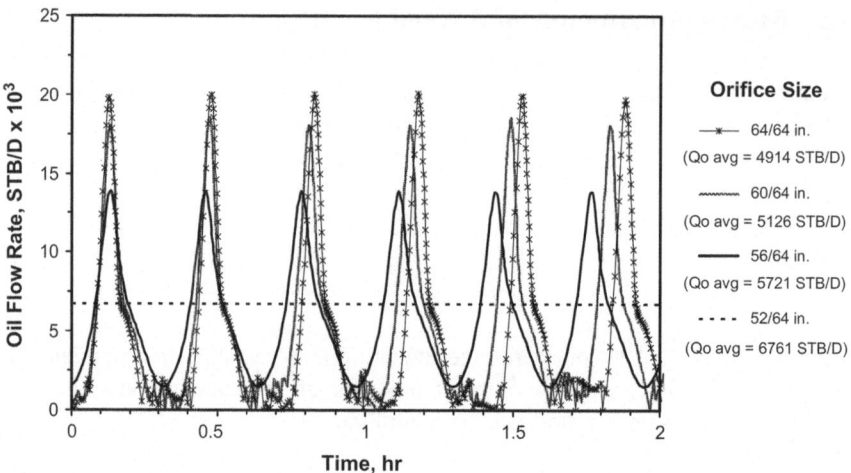

Fig. 3.2 Change in flow rate in an unstable flow (Guerrero-Sarabia and Fairuzov 2013)

Asheim (1988). He said that the instability occurs whenever the density of the fluid decreases as the pressure at the injection point increases. In this situation the lighter fluid leads to more entrance of lift gas to tubing. Thus, (3.2) is necessary to avoid the instable flow.

$$\frac{\partial \rho_i(t)}{\partial p_{ti}(t)} < 0 \qquad (3.2)$$

He said if (3.2) is satisfied the flow is stable. However, if it necessarily violated the flow it is not unstable. In this condition, it can be stable if (3.3) is satisfied.

$$\frac{\partial \rho_i(t)}{\partial p_{ti}(t)} < 0 \qquad (3.3)$$

Based on the above explanation he introduced two factors for the stability prediction, F_1 and F_2. To insure the stable flow, at least one of these factors should be more than 1. These factors are defined as below:

$$F_1 = \frac{\rho_{gsc} B_g q_{gsc}^2}{q_{Lsc}} \frac{J}{(EA_i)^2} \qquad (3.4)$$

$$F_2 = C \frac{V_t}{V_c} \frac{1}{gD} \frac{p_{ti}}{\rho_{fi} - \rho_{gi}} \frac{q_{fi} + q_{gi}}{q_{fi}(1 - F_1)} \qquad (3.5)$$

$$C = \frac{p_{ti} T_{ci} z_{ci}}{pci T_{ti} z_{ti}} \cong 1 \qquad (3.6)$$

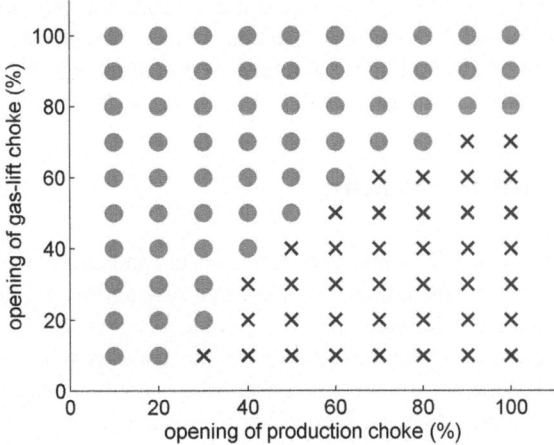

Fig. 3.3 A typical stability map based on opening of choke size, the circles show the stable region (Shao et al. 2016)

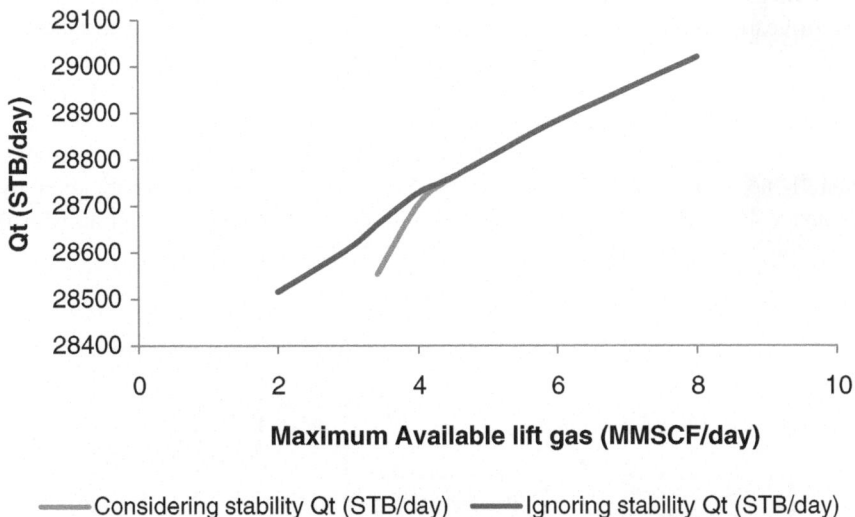

Fig. 3.4 The amount of total produced oil in cases of considering and ignoring stability in gas allocation optimization for different amounts of available lift gas in a typical problem (Mahdiani and Khamehchi 2015)

In addition to using the equation for predicting the gas lift instability, some researchers drew some maps that were based on different well and reservoir properties predicting the flow instability. Figure 3.3 shows a typical stability map. Mahdiani and Khamehchi (2015) considered this as a constraint in a gas allocation optimization problem and then compared the optimum point with the situation when it was ignored. They stated that considering this does not cause a huge production loss while it discards instability. They also indicated that while the maximum amount of available lift gas increases this difference reduces (Fig. 3.4).

3.6 Controlling the System

Usually in gas lift operations some disturbance occurs and causes the system to lose its optimal situation (Salahshoor et al. 2013). For solving this problem, controllers can help (Sharma and Glemmestad 2013). Some researchers have used control methods to ensure the maximum efficacy of the used system. Usually, in this approach as well as the normal process, a feedback is continuously returning to the operator to correct disturbances. Figure 3.5 shows the flow of information in a controlled system.

Aliev et al. (2010, 2011) considered the flow regime in the tubing as an important factor and tried to control it. He tried to stabilize the flow in the pipe by controlling the pressure and volume of injected gas. Hussein et al. (2015) considered the gas lift itself, and found it be very nonlinear and designed a relay feedback controller to control it. In some control systems some sensors are installed to continuously monitor the condition of the well and gas lift process. One of the works in this area is the research of Romer (2016). A schematic of his work is illustrated in Fig. 3.6. For more information see Romer (2016).

Control methods are very good for dealing with constraints. For this purpose, there is need to a control the oriented model. Usually these methods are given by ordinary differential equations or partial differential equations. Shao et al. (2016) used these kinds of models to consider the stability in his study.

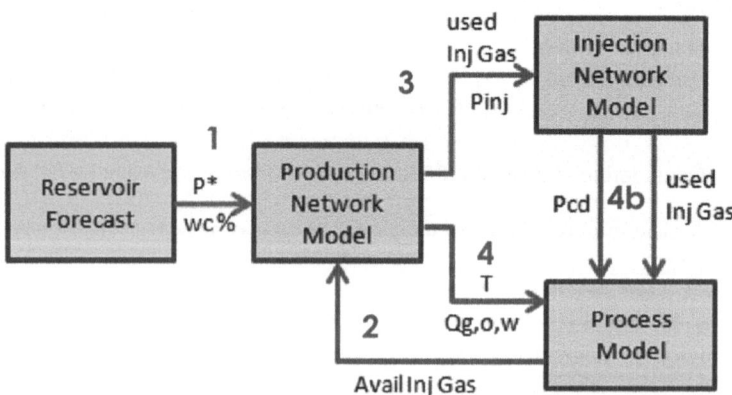

Fig. 3.5 Flow of information in a controlled gas allocation problem (Gutierrez et al. 2007)

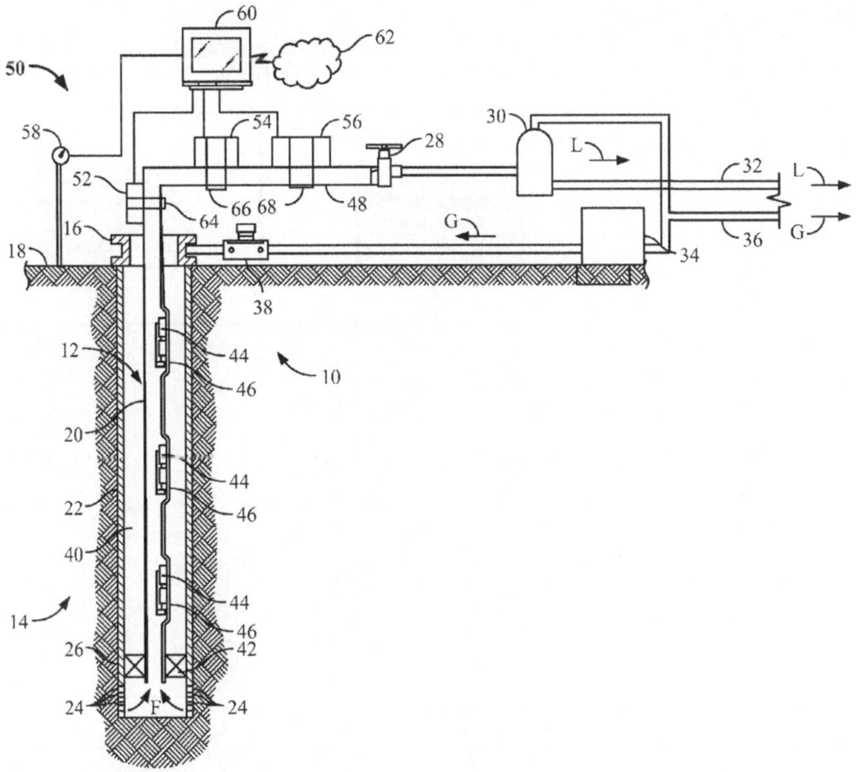

Fig. 3.6 Controlling a well (Romer 2016)

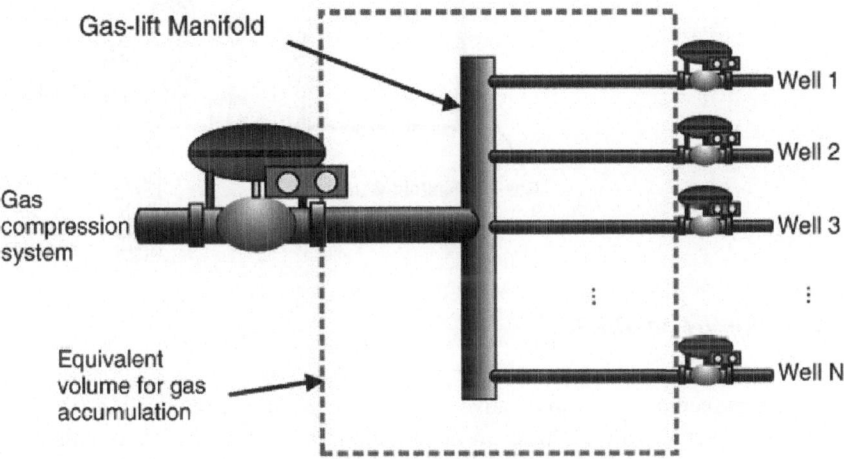

Fig. 3.7 The smart system of gas allocation in gas lift (Camponogara et al. 2010)

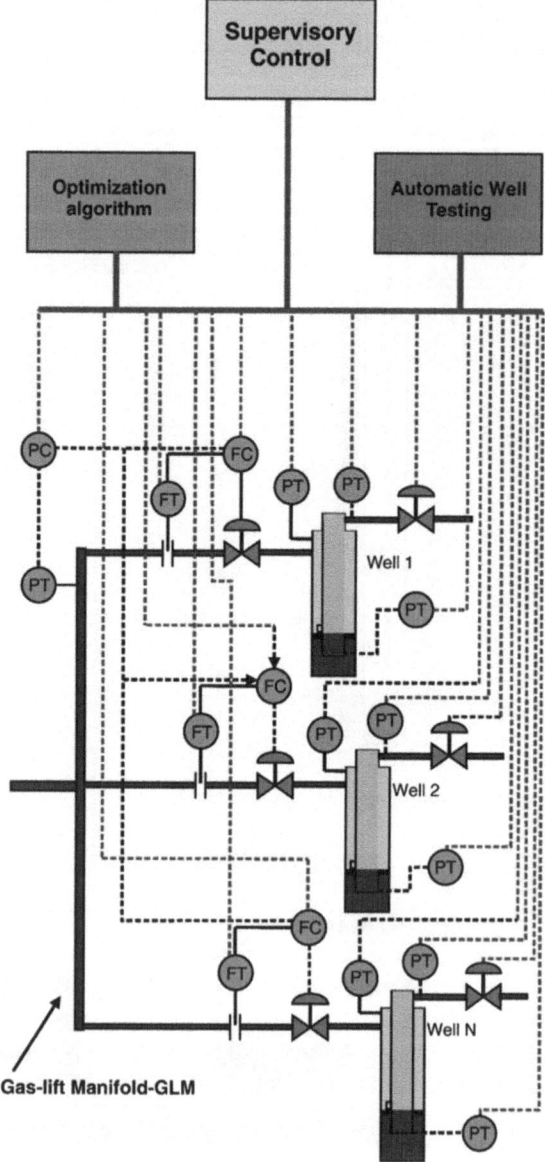

Fig. 3.8 Gas lift automation system (Camponogara et al. 2010)

3.6.1   Smart Methods

Using smart methods is an innovative way to increase the oil production (DeVries 2005; Murray et al. 2006). The smart fields can optimize the production in the long run (Williams and Webb 2007). The first use of smart field methods dates back to the 1960s. There are currently much more advanced technologies in this area that contain a design of completion methods and downhole instruments and sensors for pressure, temperature and fluid flow (Williams and Webb 2007). These data are transformed to the surface for analysis and optimization (Bogaert et al. 2004). In addition, the smart technologies can be extended to surface facilities such as the compressor and surface valves (Moises et al. 2008). Smart field development can be regarded as a method for increasing the total production which consists of sensors (Nath et al. 2006; Aref et al. 2007), data accusation (DeVries 2005) and reservoir optimization (Yeten and Jalali 2001; Yeten et al. 2004).

Camponogara et al. (2010) studied the smart field applicability in gas allocation optimization methods. Their method consisted of an algorithm that was allocating gas to some wells by considering the constraints. In this study, they assumed that all wells are equipped with pressure and temperature sensors and each well has its own controller for lift gas rate. The controller tries to assign the injection rates in order to maximize the oil production as shown in Fig. 3.7.

Another schematic of gas lift automation is shown in Fig. 3.8. As this figure shows, the system can be updated by information of the sensors (PT: pressure and temperature). Using this data from downhole and surface and analyzing them can be a method for watching the reservoir performance or some kind of well test. Based on the gained data, the system controls the flow (FC).

References

Aliev, F. A., Dzhamalbekov, M. A., & Il'yasov, M. K. (2011). Mathematical simulation and control of gas lift. *Journal of Computer and Systems Sciences International, 50*(5), 805–814.

Aliev, Fa, Il'yasov, M. K., & Nuriev, N. B. (2010). Problems of modeling and optimal stabilization of the gas-lift process. *International Applied Mechanics, 46*(6), 709–717.

Aliev, F. A., & Ismailov, N. A. (2015). Optimization problems with periodic boundary conditions and boundary control for gas-lift wells. *Journal of Mathematical Sciences, 208*(5), 467–476.

Aref, S. H., et al. (2007). Fiber optic Fabry-Perot pressure sensor with low sensitivity to temperature changes for downhole application. *Optics Communications, 269*(2), 322–330.

Asheim, H. (1988). Criteria for gas-lift stability. *Journal of Petroleum Technology, 40*(11), 1452–1456.

Bogaert, P. M., et al. (2004). Improving oil production using smart fields technology in the sf30 satellite oil development offshore malaysia. In *Offshore Technology Conference*.

Camponogara, E., et al. (2010). An automation system for gas-lifted oil wells: Model identification, control, and optimization. *Journal of Petroleum Science and Engineering, 70*(3), 157–167.

Chithra Chakra, N., et al. (2015). An innovative neural forecast of cumulative oil production from a petroleum reservoir employing higher-order neural networks (HONNs). *Journal of Petroleum Science and Engineering, 26*(3–4), 18–33.

DeVries, S. G. (2005). Production management information challenges of the digital oil field. In *SPE Annual Technical Conference and Exhibition*. Society of Petroleum Engineers.

Ebrahimi, A., & Khamehchi, E. (2016). Developing a novel workflow for natural gas lift optimization using advanced support vector machine. *Journal of Natural Gas Science and Engineering, 28*, 626–638.

Fairuzov, Y. V., et al. (2004). Stability maps for continuous gas-lift wells: A new approach to solving an old problem. In *SPE Annual Technical Conference and Exhibition* (pp. 1–9). Society of Petroleum Engineers.

Ghaedi, M., Ghotbi, C., & Aminshahidy, B. (2013). Optimization of gas allocation to a group of wells in a gas lift using an efficient ant colony algorithm (ACO). *Petroleum Science and Technology, 31*(11), 949–959.

Guerrero-Sarabia, I., & Fairuzov, Y. V. V. (2013). Linear and non-linear analysis of flow instability in gas-lift wells. *Journal of Petroleum Science and Engineering, 108*, 162–171.

Gutierrez, F. A., et al. (2007). A new approach to gas lift optimization using an integrated asset model. In *International Petroleum Technology Conference.*

Hou, J., et al. (2016). Hybrid optimization technique for cyclic steam stimulation by horizontal wells in heavy oil reservoir. *Computers & Chemical Engineering, 84*, 363–370.

Hussein, H., Al-Durra, A., & Boiko, I. (2015). Design of gain scheduling control strategy for artificial gas lift in oil production through modified relay feedback test. *Journal of the Franklin Institute, 352*(11), 5122–5144.

Khamehchi, E., Abdolhosseini, H., & Abbaspour, R. (2014). Prediction of maximum oil production by gas lift in an Iranian field using auto-designed neural network. *Academic Research Online Publisher, 2*(2), 138–150.

Mahdiani, M. R., & Khamehchi, E. (2015). Stabilizing gas lift optimization with different amounts of available lift gas. *Journal of Natural Gas Science and Engineering, 26*, 18–27.

Moises, G. V. L., Rolim, T. A., & Formigli, J. M. (2008). Gedig: Petrobras corporate program for digital integrated field management. In *Intelligent Energy Conference and Exhibition.* Society of Petroleum Engineers.

Murray, R. B., et al. (2006). Making our mature fields smarter—An industry wide position paper from the 2005 SPE Forum. In *Intelligent Energy Conference and Exhibition.* Society of Petroleum Engineers.

Nath, D. K., Finley, D. B., & Kaura, J. D. (2006). Real-time fiber-optic distributed temperature sensing (DTS)—New applications in the oilfield. In *SPE Annual Technical Conference and Exhibition.* Society of Petroleum Engineers.

Ranjan, A., Verma, S., & Singh, Y. (2015). Gas lift optimization using artificial neural network. In *SPE Middle East Oil & Gas Show and Conference.* Society of Petroleum Engineers.

Rasouli, H., Rashidi, F., & Khamehchi, E. (2011). Optimization of an integrated model to enhance oil production based on gas lift optimization under limited gas supply. *Oil Gas European Magazine, 37*(4), 199–202.

Ray, T., & Sarker, R. (2007). Genetic algorithm for solving a gas lift optimization problem. *Journal of Petroleum Science and Engineering, 59*(1–2), 84–96.

Romer. (2016). US20160053593A1.pdf.

Salahshoor, K., Zakeri, S., & Haghighat Sefat, M. (2013). Stabilization of gas-lift oil wells by a nonlinear model predictive control scheme based on adaptive neural network models. *Engineering Applications of Artificial Intelligence, 26*(8), 1902–1910.

Shao, W., Boiko, I., & Al-Durra, A. (2016). Control-oriented modeling of gas-lift system and analysis of casing-heading instability. *Journal of Natural Gas Science and Engineering, 29*, 365–381.

Sharma, R., & Glemmestad, B. (2013). On generalized reduced gradient method with multi-start and self-optimizing control structure for gas lift allocation optimization. *Journal of Process Control, 23*(8), 1129–1140.

Takács, G. (2005). *gas lift manual.* Tulsa: PennWell.

Williams, C., & Webb, T. (2007). Shell strives to make smart fields smarter. *Society of Petroleum Engineers.*

Yeten, B., & Jalali, Y. (2001). Effectiveness of intelligent completions in a multiwell development context. In *SPE Middle East Oil Show* (pp. 1–7). Society of Petroleum Engineers.

Yeten, B., et al. (2004). Decision analysis under uncertainty for smart well deployment. *Journal of Petroleum Science and Engineering, 44*(1), 175–191.

Chapter 4
Optimization Algorithms

Abstract The main part of every optimization problem is the optimizer and the gas allocation optimization problem is not an exception. There are different optimization algorithms that are applicable in these kind of problems. Generally, these algorithms are divided into two main groups of numerical and heuristic methods. Traditionally, the numerical methods were common in use. These methods such as equal slope, are based on some routine calculations or plots and their answers are absolute which means that different times of using them in a specific problem results in the same answer and finally their answer is the best possible one. However, their problem is that as the number of involved parameters increases, their degree of complexity increases unimaginably. On the other side there are the heuristic methods. These methods are random based and their different runs lead to different solutions (may be near each other). However, their advantage is that they can deal with complex problems much more effectively than numerical ones, specially, in modern problems in which the number of input parameters is large. In this chapter, the different methods with their algorithms and their mathematical equations will be discussed. Finally, in some examples the accuracy and runtime of different algorithms will be compared.

Keywords Optimization algorithms · Numerical optimization · Heuristic algorithms

4.1 Introduction

There are different types of optimization algorithms that can be used in gas allocation optimization. Generally they can be classified into two categories: numerical algorithms and heuristic ones (Jacoud et al. 2015).

© The Author(s) 2017   35
E. Khamehchi and M.R. Mahdiani, *Gas Allocation Optimization Methods in Artificial Gas Lift*, SpringerBriefs in Petroleum Geoscience & Engineering, DOI 10.1007/978-3-319-51451-2_4

4.2 Numerical Algorithms

Until some years ago using numerical methods for finding an optimum point for a gas allocation problem was a common method. These methods require an initial guess of the solution, and then the process moves in search direction d^k (see (4.1)).

$$d^k = \left(d_1^k \cdot d_2^k \cdot \ldots \cdot d_n^k\right). \tag{4.1}$$

The general form of updating the gas injection rates is as follows (Nishikiori et al. 1989):

(a) Set k = 0
(b) If the Q_g^k is optimum terminate the computation otherwise determine d^k for Q_g^k
(c) Find the step length α^k that maximizes $f(Q_g^k + \alpha^k d^k)$
(d) Set $Q_g^{k+1} = Q_g^k + \alpha^k d^k$ and set k = k + 1
(e) to (b)

There are various methods to find the search direction d^k and α^k in different steps until the optimum point is found.

4.2.1 Equal Slope Optimization

The equal slope optimization is a method for finding the best allocation. Kanu et al. (1981) expressed this in 8 steps:

Step 1 Analyze the wells and calculate the well performance for different gas liquid ratio in gas lift operation.
Step 2 Establish a relation for the production oil rate versus injection gas. These plots are called gas lift performance curve. Figure 4.1 shows a typical gas lift performance curve.
Step 3 Plot the data of Step 2 for all wells in a unique graph.
Step 4 Draw lines with various slopes tangent to each curve (as Fig. 4.2).
Step 5 At each point of Step 4 find the injection rate and production.
Step 6 Establish a relationship between slope and the injection and production rates for each well.
Step 7 Establish a relationship between slope and the injection and production rates for the whole field by calculating the equation of Step 6.
Step 8 Calculate the economic slope using Eq. (4.2):

$$m = \frac{\Delta q_L}{\Delta q_g} = \frac{C_g}{f_o P}. \tag{4.2}$$

Step 9 Use this slope and use it in Step 6.

Fig. 4.1 A typical gas lift
performance curve (Rashid
et al. 2012)

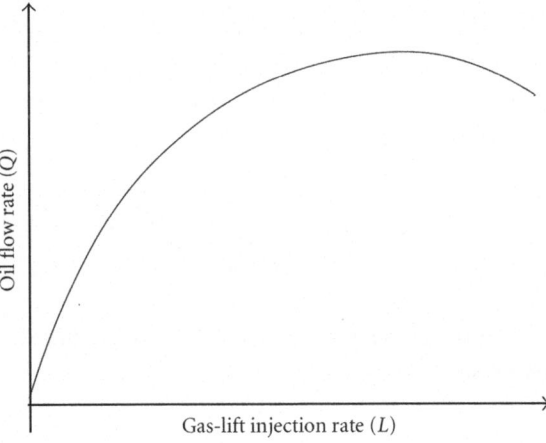

Fig. 4.2 Economic slope
(Rashid et al. 2012)

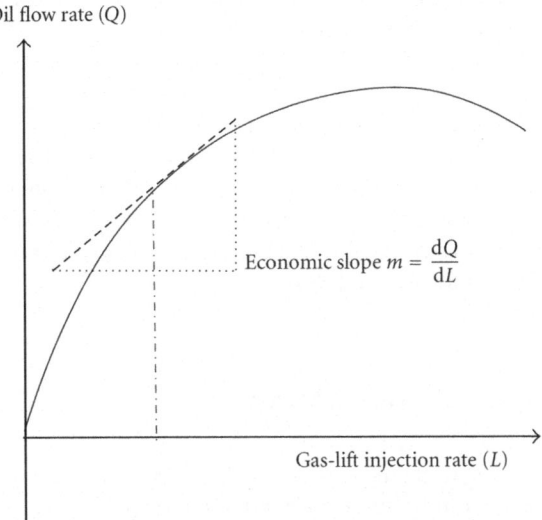

Step 10 Obtain the total injection rate by adding the optimum injection rates of
individual wells, which are gained by slopes.

4.2.2 Gradient Optimization

One of the oldest methods that sometimes was also the most common one is the
gradient or steepest ascent method (Fletcher 2013; Luenberger 1984). This function
approximates the objective (fitness) function by a first degree Taylor polynomial
(4.3):

$$f\left(Q_g^k + \delta\right) = f\left(Q_g^k\right) + \delta^T g^k. \tag{4.3}$$

In which $\delta = \alpha\, d^k$ and g^k is the gradient of "f" at Q_g^k.
For g^k see (4.4):

$$\nabla f\left(Q_g^k\right) = \left(\frac{\partial f\left(Q_g^k\right)}{\partial q_{g1}} \cdot \frac{\partial f\left(Q_g^k\right)}{\partial q_{g2}} \cdot \ldots \cdot \frac{\partial f\left(Q_g^k\right)}{\partial q_{gn}}\right)^T = g^k. \tag{4.4}$$

In this method, for increasing the total production oil rate, condition (4.5) should be satisfied:

$$d^{k^T} g^k > 0. \tag{4.5}$$

This condition is called the ascent condition. In the gradient method, the search condition is specified as (4.6):

$$d^k = g^k \tag{4.6}$$

This states that the gradient method searches in the steepest direction. This direction guarantees the finding of an optimum point for positive scalar α. However, further studies showed that this method searches linearly and thus frequently, it is slow in converging to the optimum point and this is its main disadvantage (Fletcher 2013; Luenberger 1984).

4.2.3   Newton Method

The Newton method is much faster than the gradient method. This method is derived from the second order Taylor polynomial approximation (see (4.7)).

$$f\left(Q_g^k + \delta\right) = f\left(Q_g^k\right) + \delta^T \;\nabla f\left(Q_g^k\right) + \frac{1}{2}\delta^T F\left(Q_g^k\right). \tag{4.7}$$

$F(Q_g^k)$ is the Hessian matrix of the second derivative. And δ is defined as (4.8):

$$\delta = -\left[F\left(Q_g^k\right)\right]^{-1} \;\nabla f\left(Q_g^k\right). \tag{4.8}$$

The iterative part of the equation is as (4.9):

$$Q_g^{k+1} = Q_g^k - \left[F\left(Q_g^k\right)\right]^{-1} \nabla f\left(Q_g^k\right). \tag{4.9}$$

The idea in Quasi-Newton is to define H as (4.10):

$$-\left[F\left(Q_g^k\right)\right]^{-1} = H^k \tag{4.10}$$

And for its iterative purposes (4.11) is defined as:

$$H^{k+1} = \left[H^k - \frac{H^k y^k y^{k^T} H^k}{y^{k^T} H^k y^k}\right] \gamma^k - \frac{\delta^k \delta^{k^T}}{\delta^{k^T} y^k} \tag{4.11}$$

The parameters of (4.11) are defined in (4.12)–(4.15):

$$\gamma^k = -\frac{\delta^{k^T} y^k}{y^{k^T} H^k y^k} \tag{4.12}$$

$$y^k = g^{k+1} - g^k. \tag{4.13}$$

$$\delta^k = Q_g^{k+1} - Q_g^k \tag{4.14}$$

$$d^k = H^k g^k \tag{4.15}$$

There are other mathematical methods for optimization that the interested reader can find in Rao (2009, Iqbal (2013). A lot of them have been used in gas allocation optimization. For example, Edwards et al. (1990) used numerical methods to create a model for gas allocation optimization. He considered the facilities in his model.

Dutta-Roy and Kattapuram (1997) used mixed-integer linear programming optimized gas allocation optimization. They proposed a model of wells and some surface facilities. The main idea in their work was to see the effect of interaction of wells in the result. Alarcón et al. (2002) used nonlinear constrained programming for solving the gas allocation optimization problem; He used the Nishikiori (Nishikiori et al. 1989) method, but modified that by using sequential quadratic programming. Fang and Lo (1996) used a linear programming method for solving this problem and Wang et al. (2002) used mixed integer non-linear programming to generalize the previous approaches. Camponogara and Nakashima (2006) used a recursive algorithm to solve the problem. Camponogara and de Conto (2005) used a piecewise linear method. Their model was based on mixed integer linear programming. Guyaguler and Byer (2008) used mixed-integer linear programming for solving this problem. Khishvand et al. (2015) used a nonlinear programming approach for solving this problem. In addition to the mentioned works, there are some other numerical methods for gas allocation optimization in McCracken and

Chorneyko (2006), Lo (1992), Staudtmeister and Rokahr (1997) and El-Massry and Price (1995).

The numerical methods were common for years. However, they suffered from a high complexity in the problems with a little more complexity. They were very slow when the number of parameters increased and had some big problems when dealing with constraint optimization. Thus, using them for all people in all problems was not an easy and applicable way, so some new methods were born.

4.3 Heuristic Algorithms

As the problems became more complex, the number of variables increased and using numerical methods became more tedious. In this situation, using heuristic algorithms became much more attractive (Lima Silva et al. 2015; Buitrago et al. 2016; Christensen and Bastien 2016).

In heuristic algorithms, some possible solutions are initially selected, then during some iterations (generations) this population is modified until a satisfying solution is found. There are different algorithms in this category that have been used or can be used in a gas allocation optimization problem such as: Genetic Algorithm (GA) (Ray and Sarker 2007; Ghaedi et al. 2013), Scatter Search (SS) (Chithra Chakra et al. 2013), Simulated Annealing (Raoufi et al. 2015), Tabu Search (Anon 2010), Artificial immune system (Araujo et al. 2003), Memetic Algorithm (Neri and Cotta 2012), Ant Colony Algorithm (ACO) (Ghaedi et al. 2013), Particle Swarm Optimization (PSO) (Hamedi et al. 2011; Hamedi and Khamehchi 2012), Differential Evolution (DE) (Price et al. 2006), Cross Entropy Method (CEM) (Bejan 1995), Harmony Search (HS) (Anon 2011), Bootstrap Algorithm (BA) (Slupphaug and Elgsaeter 2013), Bees Optimization (BO) (Jansen and Shoham 1994), Glowworm Swarm Optimization (GSO) (Fonseca and Fleming 1995), Bee Colony Algorithm (ABC) (Zitzler et al. 2000), Honey bee Mating Optimization (HMO) (Afshar et al. 2007), Intelligent Water Drops (IWD) (Shah-Hosseini 2009), Imperialist Competitive Algorithm (ICA) (Atashpaz-Gargari and Lucas 2007), Monkey Search (MS) (Mucherino et al. 2007), League Championship Algorithm (LCA) (Husseinzadeh Kashan 2011), Gravitational Search Algorithm (GSA) (Su and Wang 2015), Bat Algorithm (BA) (Yang 2011), Galaxy based Search Algorithm (GbSA) (Shah-Hosseini 2011), Spiral Optimization (SO) (Benasla et al. 2014), Teaching Learning Based Optimization (TLBO) (Rao et al. 2011), Krill Herd (KH) Algorithm (Gandomi and Alavi 2012), Differential Search Algorithm (DSA) (Price et al. 2006), firefly optimization (Kisi and Parmar 2016), bat optimization (Meng et al. 2015), cuckoo search (Huang et al. 2016).

As an example, Fig. 4.3 shows a pseudo code of the genetic algorithm, and other algorithms have a similar procedure.

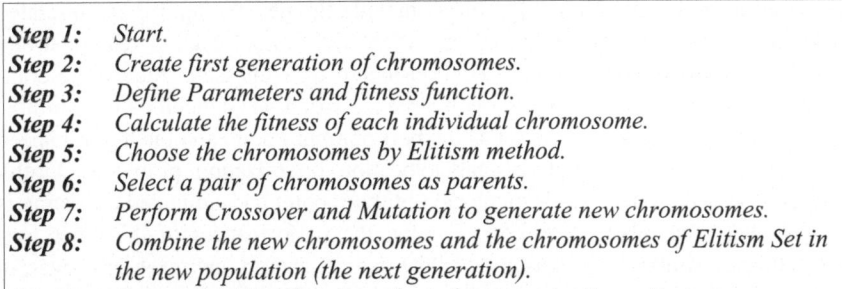

Step 1:	*Start.*
Step 2:	*Create first generation of chromosomes.*
Step 3:	*Define Parameters and fitness function.*
Step 4:	*Calculate the fitness of each individual chromosome.*
Step 5:	*Choose the chromosomes by Elitism method.*
Step 6:	*Select a pair of chromosomes as parents.*
Step 7:	*Perform Crossover and Mutation to generate new chromosomes.*
Step 8:	*Combine the new chromosomes and the chromosomes of Elitism Set in the new population (the next generation).*
Step 9:	*Repeat Step 4 to Step 8 until reaching termination criteria.*
Step 10:	*Return best solution.*

Fig. 4.3 Pseudo code of genetic algorithm (Beheshti et al. 2013)

These algorithms find the optimum solutions by step by step modification. Figure 4.4 shows the optimization process in a gas allocation optimization with heuristic algorithms.

There are some works that have used a hybrid of Heuristic algorithms for gas allocation optimization. Zerafat et al. (2009) and Khamehchi et al. (2009) used both the genetic algorithm and ant colony and Ghaedi et al. (2013) used a hybrid of the genetic algorithm for solving this optimization problem. Rasouli et al. (2015) used a hybrid of the genetic algorithm and neural network and created a real-time optimization. Mahdiani and Khamehchi (2015) compared the genetic algorithm and a hybrid of the genetic algorithm and quasi-Newton for solving the problem and said using the hybrid was a more efficient method. Mahdiani (2013) in his M.Sc. thesis compared some of the most common heuristic algorithms for gas allocation

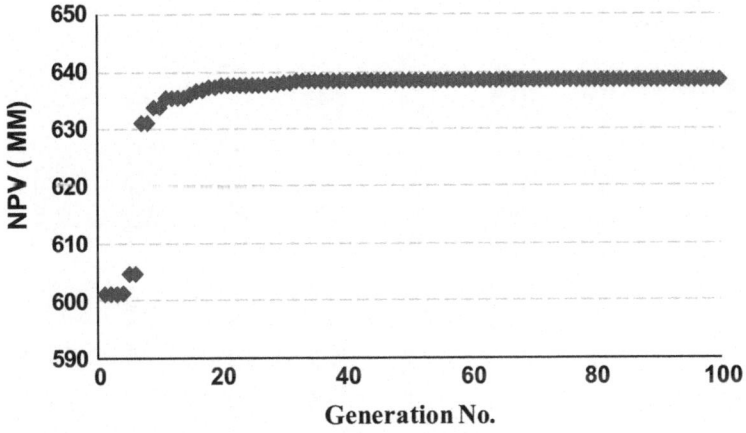

Fig. 4.4 Using heuristic algorithm to maximize the NPV in a gas allocation optimization (Mahmudi and Sadeghi 2013)

optimization problems. These algorithms include the genetic algorithm, simulated annealing, particle swarm optimization, differential search, cuckoo search, firefly optimization and harmony search. He considered different case studies and compared their optimum points and the convergence speed. He concluded that in most cases particle swarm optimization has the best optimum point and the highest speed and is highly recommended for gas allocation optimization problems. Firefly optimization occasionally leads to a local optimum point and simulated annealing is often slower than other algorithms. Finally, the performances of the other four algorithms are similar but not as good as the particle swarm optimizer. However, in some way their results can be accepted. During his studies he observed that in most cases firefly optimization found a local optimum point. But on the other hand, the rate of optimum point improvement in different iterations is very fast. After summarizing the result of the performance of different algorithms he concluded that the simulated annealing can find a good optimum point but its problem is that this algorithm is very slow. It seems that if the problem was first optimized by another algorithm and then the found optimum point was used as the start point of the simulated annealing the resulted point could have a very good total production oil rate. In one case he injected 18 MMscf/d gas to 20 different wells by various heuristic algorithms and then he compared their total oil production. Figure 4.5 shows the amount of total oil production.

For comparing the speed of these algorithms he did not compare the runtime of the optimizers, because it depends on the used computer and its internal hardware and software configuration. Instead he compared the number of fitness function evaluation. Figure 4.6 shows the number of fitness function evaluation of different algorithms.

In most of the considered cases Mahdiani saw the huge number of fitness function evaluation of the simulated annealing in comparison to other algorithms.

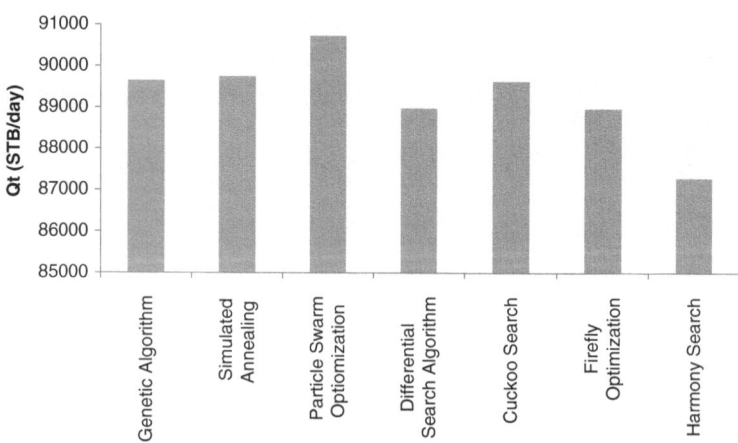

Fig. 4.5 The comparison of the total oil production of allocating 18 MMscf gas to 20 wells by different heuristic algorithms (Mahdiani 2013)

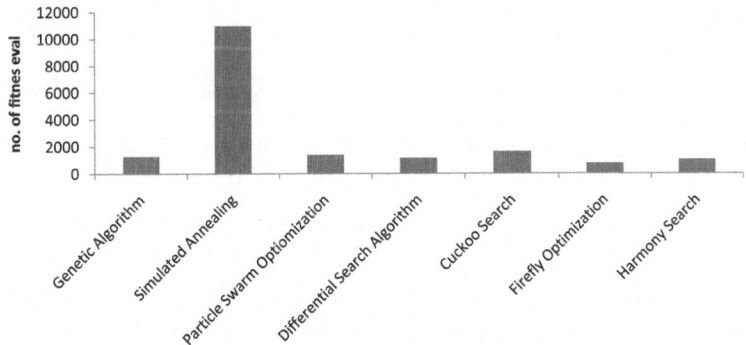

Fig. 4.6 The comparison of the total amount of fitness function evaluation of the heuristic algorithms in allocating 18 MMscf of gas to 20 wells (Mahdiani 2013)

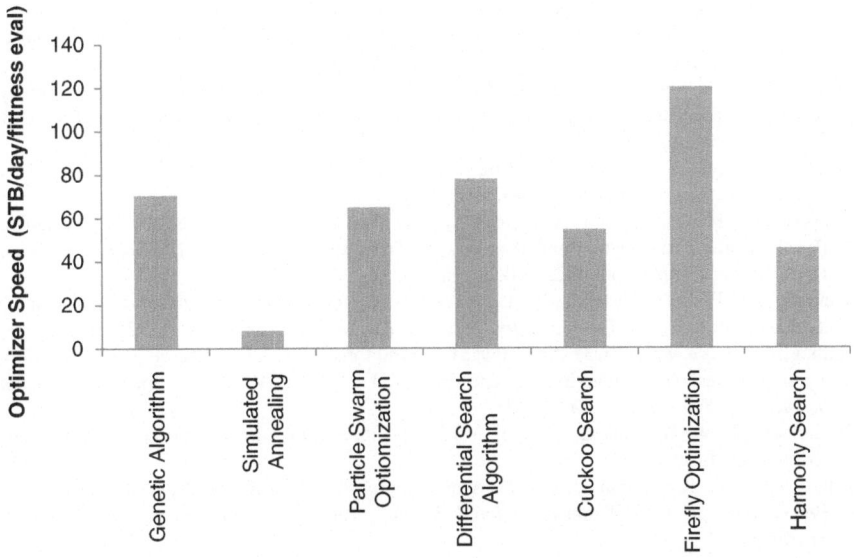

Fig. 4.7 The comparison of the optimizer speed of the heuristic algorithms in allocating 18 MMscf of gas to 20 wells (Mahdiani 2013)

In addition to the above factors, he considered another factor called optimizer speed. This showed the average amount of fitness function improvement by the number of fitness function evaluation (Fig. 4.7).

Mahdiani also changed the number of wells and maximum amount of available lift gas and repeated his calculation to see the application of the optimization algorithms in different conditions.

References

Afshar, A., et al. (2007). Honey-bee mating optimization (HBMO) algorithm for optimal reservoir operation. *Journal of the Franklin Institute, 344*(5), 452–462.

Alarcón, G. A., Torres, C. F., & Gómez, L. E. (2002). Global optimization of gas allocation to a group of wells in artificial lift using nonlinear constrained programming. *Journal of Energy Resources Technology, 124*(4), 262.

Anon, (2010). *Soft computing for recognition based on biometrics.* Berlin: Springer.

Anon. (2011). *Computational optimization and applications in engineering and industry.* Berlin: Springer Science & Business Media.

Araujo, M., Aguilar, J., & Aponte, H. (2003). Fault detection system in gas lift well based on artificial immune system. In *Proceedings of the International Joint Conference on Neural Networks* (pp. 1673–1677). IEEE.

Atashpaz-Gargari, E., & Lucas, C. (2007). Imperialist competitive algorithm: An algorithm for optimization inspired by imperialistic competition. In *2007 IEEE Congress on Evolutionary Computation* (pp. 4661–4667). IEEE.

Beheshti, Z., et al. (2013). A review of population-based meta-heuristic algorithm A Review of Population-based Meta-Heuristic Algorithm. *International Journal of Advances in Soft Computing and Its Applications, 5,* 1–35.

Bejan, A. (1995). *Entropy generation minimization: The method of thermodynamic optimization of finite-size systems and finite-time processes.* New York: CRC Press.

Benasla, L., Belmadani, A., & Rahli, M. (2014). Spiral optimization algorithm for solving combined economic and emission dispatch. *International Journal of Electrical Power & Energy Systems, 62,* 163–174.

Buitrago, M., et al. (2016). Designing construction processes in buildings by heuristic optimization. *Engineering Structures, 111,* 1–10.

Camponogara, E., & de Conto, A. M. (2005). Gas-lift allocation under precedence constraints: Piecewise-linear formulation and K-covers. In *Proceedings of the 44th IEEE Conference on Decision and Control* (pp. 4422–4427). IEEE.

Camponogara, E., & Nakashima, P. H. R. (2006). Solving a gas-lift optimization problem by dynamic programming. *European Journal of Operational Research 174*(2), 1220–1246.

Chithra Chakra, N., et al. (2013). An innovative neural forecast of cumulative oil production from a petroleum reservoir employing higher-order neural networks (HONNs). *Journal of Petroleum Science and Engineering, 106,* 18–33.

Christensen, J., & Bastien, C. (2016). *Nonlinear optimization of vehicle safety structures.* Amsterdam: Elsevier.

Dutta-Roy, K., & Kattapuram, J. (1997). A new approach to gas-lift allocation optimization. In *Proceedings of SPE Western Regional Meeting* (pp. 685–691). Society of Petroleum Engineers.

Edwards, R., Marshall, D. L., & Wade, K. C. (1990). A gas-lift optimization and allocation model for manifolded subsea wells. In *European Petroleum Conference*. Society of Petroleum Engineers.

El-Massry, Y. A.-W., & Price, A. D. (1995). Development of a network and gas lift allocation model for production optimization in the Ras Budran field. In *Proceedings of Middle East Oil Show*. Society of Petroleum Engineers.

Fang, W. Y., & Lo, K. K. (1996). A generalized well management scheme for reservoir simulation. *SPE Reservoir Engineering, 11*(02), 116–120.

Fletcher, R. (2013). *Practical methods of optimization, Second Edition—Fletcher—Wiley Online Library.* NewYork: Wiley.

Fonseca, C. M., & Fleming, P. J. (1995). An overview of evolutionary algorithms in multiobjective optimization. *Evolutionary Computation, 3*(1), 1–16.

Gandomi, A. H., & Alavi, A. H. (2012). Krill herd: A new bio-inspired optimization algorithm. *Communications in Nonlinear Science and Numerical Simulation, 17*(12), 4831–4845.

Ghaedi, M., Ghotbi, C., & Aminshahidy, B. (2013). Optimization of gas allocation to a group of wells in a gas lift using an efficient ant colony algorithm (ACO). *Petroleum Science and Technology, 31*(11), 949–959.

Guyaguler, B., & Byer, T. J. (2008). A new rate-allocation-optimization framework. *SPE Production & Operations, 23*(04), 448–457.

Hamedi, H., & Khamehchi, E. (2012). A nonlinear approach to gas lift allocation optimization with operational constraints using particle swarm optimization and a penalty function. *Petroleum Science and Technology, 30*(8), 775–785.

Hamedi, H., Rashidi, F., & Khamehchi, E. (2011). A novel approach to the gas-lift allocation optimization problem. *Petroleum Science and Technology, 29*(4), 418–427.

Huang, L., et al. (2016). Chaos-enhanced Cuckoo search optimization algorithms for global optimization. *Applied Mathematical Modelling, 40,* 3860–3875.

Husseinzadeh Kashan, A. (2011). An efficient algorithm for constrained global optimization and application to mechanical engineering design: League championship algorithm (LCA). *Computer-Aided Design, 43*(12), 1769–1792.

Iqbal, K. (2013). *Fundamental engineering optimization methods,* ISBN 978-87-403-0489-3, 1st edition.

Jacoud, A., et al. (2015). Modelling and extremum seeking control of gas lifted oil wells. *IFAC-PapersOnLine, 48*(2012), 21–26.

Jansen, F. E., & Shoham, O. (1994). Methods for eliminating pipeline-riser flow instabilities. In *SPE Western Regional Meeting.*

Kanu, E. P., Mach, J., & Brown, K. E. (1981). Economic approach to oil production and gas allocation in continuous gas lift (includes associated papers 10858 and 10865). *Journal of Petroleum Technology, 33*(10), 1887–1892.

Khamehchi, E., et al. (2009). Continuous gas lift optimization with a novel genetic algorithm. *Australian Journal of Basic and Applied Sciences, 1*(4), 587–594.

Khishvand, M., Khamehchi, E., & Nokandeh, N. R. (2015). A nonlinear programming approach to gas lift allocation optimization. *Energy Sources, Part A: Recovery, Utilization, and Environmental Effects, 37*(5), 453–461.

Kisi, O., & Parmar, K. S. (2016). Application of least square support vector machine and multivariate adaptive regression spline models in long term prediction of river water pollution. *Journal of Hydrology, 534,* 104–112.

Lima Silva, T., et al. (2015). Modeling of flow splitting for production optimization in offshore gas-lifted oil fields: Simulation validation and applications. *Journal of Petroleum Science and Engineering, 128,* 86–97.

Lo, K. K. (1992). *Optimum lift-gas allocations under multiple production constraints.*

Luenberger, D. G. (1984). *Linear and nonlinear programming.* New York: Addison-Wesley.

Mahdiani, M. R. (2013). *Hydrocarbon fields development optimization, emphasizing on gas lift stability.* Amirkabir University of Technology.

Mahdiani, M. R., & Khamehchi, E. (2015). Preventing instability phenomenon in gas-lift optimization. *Iranian Journal of Oil & Gas Science and Technology, 4*(1), 49–65.

Mahmudi, M., & Sadeghi, M. T. (2013). The optimization of continuous gas lift process using an integrated compositional model. *Journal of Petroleum Science and Engineering, 108,* 321–327.

McCracken, M., & Chorneyko, D. M. (2006). Rate allocation using permanent downhole pressures. In *SPE Annual Technical Conference and Exhibition.* Society of Petroleum Engineers.

Meng, X. B., et al. (2015). A novel bat algorithm with habitat selection and Doppler effect in echoes for optimization. *Expert Systems with Applications, 42*(17–18), 6350–6364.

Mucherino, A., et al. (2007). Monkey search: a novel metaheuristic search for global optimization. In *AIP Conference Proceedings* (pp. 162–173). AIP.

Neri, F., & Cotta, C. (2012). Memetic algorithms and memetic computing optimization: A literature review. *Swarm and Evolutionary Computation, 2,* 1–14.

Nishikiori, N., et al. (1989). An improved method for gas lift allocation optimization. In *SPE Annual Technical Conference and Exhibition*. Society of Petroleum Engineers.

Price, K., Storn, R. M., & Lampinen, J. A. (2006). *Differential evolution: A practical approach to global optimization*. Berlin: Springer Science & Business Media.

Rao, S. S., & Rao, S. S. (2009). *Engineering optimization: Theory and practice*. John Wiley & Sons

Rao, R. V., Savsani, V. J., & Vakharia, D. P. (2011). Teaching–learning-based optimization: A novel method for constrained mechanical design optimization problems. *Computer-Aided Design, 43*(3), 303–315.

Raoufi, M. H., Farasat, A., & Mohammadifard, M. (2015). Application of simulated annealing optimization algorithm to optimal operation of intelligent well completions in an offshore oil reservoir. *Journal of Petroleum Exploration and Production Technology, 5*(3), 327–338.

Rashid, K., Bailey, W., & Couët, B. (2012). A survey of methods for gas-lift optimization. *Modelling and Simulation in Engineering*, Vol. 24.

Rasouli, E., Karimi, B., & Khamehchi, E. (2015). A surrogate integrated production modeling approach to long-term gas-lift allocation optimization. *Chemical Engineering Communications, 202*, 647–654.

Ray, T., & Sarker, R. (2007). Genetic algorithm for solving a gas lift optimization problem. *Journal of Petroleum Science and Engineering, 59*(1–2), 84–96.

Shah-Hosseini, H. (2009). The intelligent water drops algorithm: a nature-inspired swarm-based optimization algorithm. *International Journal of Bio-Inspired Computation, 1*(1–2), 71–79.

Shah-Hosseini, H. (2011). Principal components analysis by the galaxy-based search algorithm: a novel metaheuristic for continuous optimisation. *International Journal of Computational Science and Engineering, 6*(1–2), 132–140.

Slupphaug, O., & Elgsaeter, S. (2013). *Method for prediction in an oil/gas production system*. U.S. Patent No. 8,380,475.

Staudtmeister, K., & Rokahr, R. B. (1997). Rock mechanical design of storage caverns for natural gas in rock salt mass. *International Journal of Rock Mechanics and Mining Sciences, 34*(3–4), 301–313.

Su, Z., & Wang, H. (2015). A novel robust hybrid gravitational search algorithm for reusable launch vehicle approach and landing trajectory optimization. *Neurocomputing, 162*, 116–127.

Wang, P., Litvak, M., & Aziz, K. (2002). Optimization of production operations in petroleum fields. In *SPE Annual Technical Conference and Exhibition* (pp. 1–12). Society of Petroleum Engineers.

Yang, X.-S. (2011). Bat algorithm for multi-objective optimisation. *International Journal of Bio-Inspired Computation, 3*(5), 267–274.

Zerafat, M. M., Ayatollahi, S., & Roosta, A. A. (2009). Genetic algorithms and ant colony approach for gas-lift allocation optimization. *Journal of the Japan Petroleum Institute, 52*(3), 102–107.

Zitzler, E., Deb, K., & Thiele, L. (2000). Comparison of multiobjective evolutionary algorithms: empirical results. *Evolutionary Computation, 8*(2), 173–195.